Coastal waders and wildfowl in winter

Coastal waders and wildfowl in winter

Edited by

P. R. EVANS

J. D. GOSS-CUSTARD

W. G. HALE

for the British Ornithologists' Union

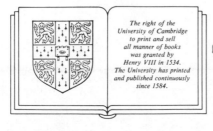

The right of the
University of Cambridge
to print and sell
all manner of books
was granted by
Henry VIII in 1534.
The University has printed
and published continuously
since 1584.

CAMBRIDGE UNIVERSITY PRESS

Cambridge

London New York New Rochelle

Melbourne Sydney

Published by the Press Syndicate of the University of Cambridge
The Pitt Building, Trumpington Street, Cambridge CB2 1RP
32 East 57th Street, New York, NY 10022, USA
296 Beaconsfield Parade, Middle Park, Melbourne 3206, Australia

First published 1984

Printed in Great Britain at the Pitman Press, Bath

Library of Congress catalogue card number: 83–17149

British Library Cataloguing in Publication Data
Coastal waders and wildfowl in winter
  1. Charadriiformes.
  I. Evans, P. R. II. Goss-Custard, J. D.
  III. Hale, W. G.
  598'. 33 QL696.C4
  ISBN 0 521 25928 2

Contents

Preface   vii

List of contributors   ix

Part 1   *Bird populations: the influence of food resources on the use of feeding areas*   1

Introduction   3
P. R. Evans

1   Coastal birds: numbers in relation to food resources   8
P. R. Evans & P. J. Dugan

2   Balancing the budget: measuring the energy intake and requirements of shorebirds in the field   29
M. W. Pienkowski, P. N. Ferns, N. C. Davidson & D. H. Worrall

3   Relations between the distribution of waders and the intertidal benthic fauna of the Oosterschelde, Netherlands   57
P. Meire & E. Kuyken

4   How Oystercatchers and Curlews successively deplete clams   69
L. Zwarts & J. Wanink

5   Waterfowl movements in relation to food stocks   84
M. R. van Eerden

6   Diving duck populations in relation to their food supplies   101
O. Pehrsson

Part 2   *Bird populations: social behaviour and the use of feeding areas*   117

Introduction   119
J. D. Goss-Custard

7   Why do birds roost communally?   123
R. C. Ydenberg & H. H. Th. Prins

8   The unsociable plover – use of intertidal areas by Grey Plovers   140
D. J. Townshend, P. J. Dugan & M. W. Pienkowski

9   Age-related distribution of Dunlin in the Dutch Wadden Sea   160
T. M. van der Have, E. Nieboer & G. C. Boere

10   Differences in quality of roosting flocks of Oystercatchers   177
C. Swennen

11   Feeding ecology, winter mortality and the population dynamics
      of Oystercatchers on the Exe estuary   190
      J. D. Goss-Custard & S. E. A. le V. dit Durell

**Part 3   *The significance of specific areas on the Palaearctic–African
      migration routes of waders***   209

      Introduction   211
      W. G. Hale

12   The Danish Wadden Sea   214
      K. Laursen & J. Frikke

13   The German Wadden Sea   224
      P. Prokosch

14   The Dutch Wadden Sea   238
      W. J. Wolff & C. J. Smit

15   The Dutch Delta Area   253
      R. J. Leewis, H. J. M. Baptist & P. L. Meininger

16   The British Isles   261
      P. R. Evans

17   The Atlantic coast of Morocco   276
      M. Kersten & C. J. Smit

18   The Banc d'Arguin (Mauritania)   293
      M. Engelmoer, T. Piersma, W. Altenburg & R. Mes

19   The changing face of European wintering areas   311
      W. G. Hale

      Index   324

Preface

Although information on the foods and habitats used by wildfowl, particularly by so-called 'quarry species' that may be hunted, has been collected steadily over many decades in this century, it was not until the late 1960s that a number of ornithologists began to study in detail the feeding ecology of coastal waders. In some cases, this was prompted by a clash of interests between shorebirds and men seeking to harvest the same prey (e.g. Oystercatchers taking the cockle *Cardium edule*); in others by study of waders as merely one component amongst the whole range of predators in an estuarine ecosystem; and in others by the need for information on the requirements of waders in coastal habitats threatened by reclamation. Accordingly, one of us (P.R.E.) organized a discussion meeting in Durham in 1972 to enable wader biologists to exchange information and techniques from their respective studies. Further meetings were held in Liverpool (organized by W.G.H.) in 1975, Durham in 1977, Liverpool in 1979 and on the island of Texel in the Netherlands in 1981 (organized by R. H. Drent, W. J. Wolff, C. Smit and C. Swennen). At an early stage it was realized that there were considerable advantages to be gained by including contributions to the discussions from research workers studying coastal wildfowl, particularly those species whose foods were subjected (like those of waders) to periodic immerson by the tides. In 1977 the 'working group' studying feeding ecology and behaviour of shorebirds became affiliated to the Feeding Ecology Research Group of the International Waterfowl Research Bureau. The Texel meeting provided the opportunity for European ornithologists studying coastal wildfowl and waders (those using both intertidal and saltmarsh habitats) to exchange views on three major topics, which form the three sections of this book. The chapters that follow provide a mixture of reviews, overviews of long-term studies, and more specific reports, all written especially for this book. The material from most of the research projects has been or will be published in detail in the scientific journals. It is hoped that this book will be of value to students of animal ecology and behaviour as well as to ornithologists interested in the conservation of the birds of coastal habitats.

In conclusion, we wish to record the valuable contributions made by Dr R. H. Drent to the discussion meetings and the overriding influence he has had on the progress to publication of this book. We are also grateful to Dr Bernard Stonehouse for advice and encouragement during the preparation of the manuscripts, to Dr N. C. Davidson and S. P. Carter for preparing several sets of figures and to the editorial staff of Cambridge University Press for their assistance.

P.R.E.

J.D.G.-C.

W.G.H.

Contributors

Altenburg, W. Zoological Laboratory, University of Groningen, Kerklaan 30, 9751 NN Haren, Netherlands

Baptist, H. J. M. Rijkswaterstaat, Deltadienst, hoofdafd Milieu en Inrichting, PO Box 43g, 4330 AK Middelburg, Netherlands

Boere, G. C. Department of Nature Conservation, PO Box 20020, 3505 LA Utrecht, Netherlands

Davidson, N. C. Department of Zoology, University of Durham, South Road, Durham DH1 3LE, UK

Dugan, P. J. Station Biologique de la Tour du Valat, Le Sambuc, 13200 Arles, France

Durell, S. E. A. le V. dit Institute of Terrestrial Ecology, Furzebrook Research Station, Wareham, Dorset BH20 5AS, UK

van Eerden, M. R. Rijksdienst voor die Ijsselmeer polders, PO Box 600, 8200 AP Lelystad, Netherlands

Engelmoer, M. Zoological Laboratory, University of Groningen, Kerklaan 30, 9751 NN Haren, Netherlands

Evans, P. R. Department of Zoology, University of Durham, South Road, Durham DH1 3LE, UK

Ferns, P. N. Department of Zoology, University College, PO Box 78, Cardiff CF1 1XL, UK

Frikke, J, Game Biology Station, Kalø, 8410 Rønde, Denmark

Goss-Custard, J. D. Institute of Terrestrial Ecology, Furzebrook Research Station, Wareham, Dorset BH20 5AS, UK

Hale, W. G. Department of Biology, Liverpool Polytechnic, Byrom Street, Liverpool L3 3AF, UK

Kersten, M. Oliemulderstraat 55, 9724 JD Groningen, Netherlands

Kuyken, E. Laboratorium voor Oecologie der Dieren, Zoögeografie en Natuurbehoud, Rijksuniversiteit Gent, 9000-GENT, Belgium

Laursen, K. Game Biology Station, Kalø, 8410 Rønde, Denmark

Leewis, R. J. Rijkswaterstaat, Deltadienst, hoofdafd Milieu en Inrichting, PO Box 43g, 4330 AK Middelburg, Netherlands

Meininger, P. L. Rijkswaterstaat, Deltadienst, hoofdafd Milieu en Inrichting, PO Box 43g, 4330 AK Middleburg, Netherlands

Meire, P. Laboratorium voor Oecologie der Dieren, Zoögeografie en Natuurbehoud, Rijksuniversiteit Gent, 9000-GENT, Belgium

Mes, R. Zoological Laboratory, University of Groningen, Kirklaan 30, 9751 NN Haren, Netherlands

Nieboer, E. Department of Biosystematics, Vrije Universiteit, Amsterdam, de Boelelaan 1087, 1007 MC Amsterdam, Netherlands

Pehrsson, O. Department of Zoology, University of Göteborg, Box 250 59, S-400 31 Göteborg, Sweden

Pienkowski, M. W. Department of Zoology, University of Durham, South Road, Durham DH1 3LE, UK

Piersma, T. Zoological Laboratory, University of Groningen, Kirklaan 30, 9751 NN Haren, Netherlands

Prins, H. H. Th. Large Animal Research Group, Department of Zoology, 34A Storey's Way, Cambridge CB3 0DT, UK

Prokosch, P. Staatliche Vogelschutzwarte Schleswig–Holstein, Olshausenstrasse 40–60, 2300 Kiel 1, Federal Republic of Germany

Smit, C. J. Research Institute for Nature Management, PO Box 59, 1790 AB Den Burg, Texel, Netherlands

Swennen, C. Netherlands Institute for Sea Research, PO Box 59, 1790 AB Den Burg, Texel, Netherlands

Townshend, D. J. Department of Zoology, University of Durham, South Road, Durham DH1 3LE, UK

Van der Have, T. M. Department of Biosystematics, Vrije Universiteit Amsterdam, de Boelelaan 1087, 1007 MC Amsterdam, Netherlands

Wanink, J. Zoological Laboratory, University of Groningen, PO Box 14, 9751 AA Haren, Netherlands

Wolff, W. J. Research Institute for Nature Management, PO Box 59, 1790 AB Den Burg, Texel, Netherlands

Worrall, D. H. Department of Zoology, University College, PO Box 78, Cardiff CF1 1XL, UK

Ydenberg, R. C. Edward Grey Institute of Field Ornithology, Department of Zoology, South Parks Road, Oxford OX1 3PS, UK

Zwarts, L. Rijksdienst voor die Ijsselmeerpolders, PO Box 600, 8200 AP Lelystad, Netherlands

PART 1

Bird populations: the influence of food resources on the use of feeding areas

Introduction

P. R. EVANS

A long tradition exists in avian population studies of monitoring changes in numbers or densities of birds, with rather little attention being paid to the size, biomass or quality of the individuals involved. Instead, interest has been focussed upon the maximum number of birds using a location, often spoken of as its 'carrying capacity'. This approach also finds its way into many arguments concerning the protection or management of habitats for birds: an area is considered good if it supports large numbers of birds at some time of year, poor if it does not. It is a major theme of this book that such attitudes are inappropriate when considering wintering shorebird and wildfowl populations. In particular we shall avoid the use of the term 'carrying capacity', in spite of its current popularity amongst ornithologists when discussing the numbers of birds found on estuaries. Although the term currently has many meanings, none is of particular value in the investigations of the problems discussed here. Thus theoretical population ecologists use the phrase to indicate the level of abundance (density) around which a population fluctuates in the long term; a level which is determined by the relationships between birth rate, death rate and density (Fig. 1). However, the population under consideration is normally considered to be closed, unaffected by fluctuations in numbers of other species in the area, and composed of identical individuals. Although this definition of 'carrying capacity' may perhaps be usefully applied to the equilibrium density of a breeding population, its relevance to coastal birds in the non-breeding season, when there may be considerable turnover of individuals at any one site, as well as highly socially structured populations, is extremely doubtful.

In wildlife management two definitions for the same term have been used, depending on the degree of interference which man can impose.

Considering those species which man may wish to harvest for food (e.g. large mammalian herbivores), the carrying capacity has been defined as the size (biomass) of a population that can be supported on a long-term, steady-state basis without detriment to the non-renewable resources of a given area or its productivity. In cases where man can alter not only the death rate but also the birth or immigration rate (as in fish populations), carrying capacity is equated with the maximum stocking density. Here again it is measured in terms of biomass density, not numerical density, since in fish, survival is possible at high densities at the expense of growth.

The 'stocking density' concept is not particularly helpful when considered in relation to coastal birds, since it is defined in relation to the effect that the predators or herbivores have on their food resources ('biomass … supported without detriment to non-renewable resources') rather than vice versa. Secondly, it is not obvious that unmanaged populations should come to be controlled naturally at maximum stocking density levels, nor how to translate biomass levels into numbers of animals. Thirdly, it should be noted that the food resources of most coastal waders and wildfowl in temperate areas are not renewed during the winter months, so that discussion should focus on the total bird-days of use that can be obtained from an area, as well as on the number of birds that are supported by the area, or could be supported at maximum at any one moment during the winter. Finally, I suggest that the conventional assumptions made in studies of the numbers of migrant

Fig. 1. A simple graphical model showing how equilibrium population density ('carrying capacity') may be determined.

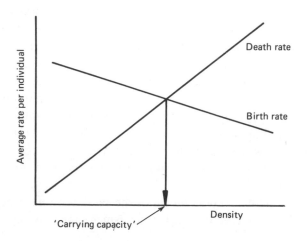

birds present in an estuary ignore several important points. For example, viewed from the population level, it is often assumed that in winter a migrant species is spread over all areas of the habitats it is adapted to use during the non-breeding season. Yet suitable areas farther from the breeding grounds may not be occupied, because a sufficient number of suitable areas is available nearer to the breeding areas. Perhaps even those wintering areas closer to the breeding grounds do not contain nearly as many birds as might be able to survive there, given the food resources they contain and the demands made upon them by each bird. Perhaps only suitable feeding areas within milder climatic regimes are 'full'. These points can be understood more easily if viewed from the context of the individual bird, whose patterns of behaviour have evolved to enhance its survival, in good nutritional condition, from one breeding season to the next. In other words an individual may not use a wintering area in which it *might* be able to survive occasionally, if, by moving further, its chances of survival are better.

In view of these uncertainties, which cast doubt on the value of the question 'how many individuals *can* an area support', studies of waders and wildfowl wintering on the coast have recently been concerned more with how individual birds attempt to ensure their survival and maintenance of good nutritional condition. In Chapter 1, Evans & Dugan review the problems which beset coastal wildfowl, and more particularly waders, in their use of intertidal food stocks during the winter months. They highlight the importance of seasonal changes in the percentage, as well as density, of invertebrate prey that are available to their shorebird predators. They point out that, if birds wish to stay in one feeding site from autumn onwards, they must choose sites when they first arrive that contain many times the density of available prey necessary to provide an adequate rate of feeding then. This ensures that when prey become least available (i.e. accessible and detectable) in mid-winter, adequate feeding rates are still possible. The published and unpublished studies that are reviewed in this chapter suggest that birds normally occur at highest densities where the density of their prey is highest, but do not answer the crucial question of how close to maximum supportable bird densities these are.

The second chapter introduces the problems encountered, and the techniques by which these may be overcome, in the study of relationships between birds and their resources. Any such study requires measurement of food requirements and rates of food intake. However, these are not easy to achieve in the field, partly because many waders and wildfowl feed at night, and partly because many feeding movements occur too quickly to be recorded by a human observer. Laboratory studies are not satisfactory substitutes for field studies of waders at present, since gathering of food by waders in the field may require as much energy again as that required for maintenance in caged conditions, and the costs of foraging movements have not been measured in captivity. Hence Pienkowski, Ferns, Davidson & Worral emphasize techniques for quantifying food intake of waders under field conditions (techniques for wildfowl are outlined later, in Chapter 5 by van Eerden).

The remainder of the first part of the book presents the results of four field studies investigating how shorebirds and wildfowl use areas of intertidal land in order to meet their energy requirements. In Chapter 3 Meire & Kuyken describe their studies of Oystercatchers (*Haematopus ostralegus*) and other waders in the Oosterschelde estuary in the southern Netherlands. Within the specialized habitat of the mussel (*Mytilus edulis*) beds at low tidal levels, they found that the first Oystercatchers to arrive in autumn settled on the areas of highest biomass density of mussels, but that later in the year they settled chiefly in poorer feeding sites. By December, the birds were evenly distributed over the whole mussel bed, irrespective of spatial variations in mussel density.

However, very few invertebrate prey species are as sedentary, conspicuous and available to predators as is the mussel. And even mussel beds change, if not their location, at least their age structure (and therefore proportion of different sizes available) from year to year, according to the position and success of settlement of the young, known as spat. Irregular spatfall is characteristic of many coastal bivalves, which in adult form can be quite long-lived animals (5–15 years). In Chapter 4 Zwarts & Wanink explore the implications of this, as exemplified in the edible clam *Mya arenaria*, which provides food for several shorebirds as it passes through certain size (= age) ranges. They show that Oystercatchers select smaller clams than Curlew (*Numenius arquata*) and discuss why.

In contrast to waders, most grazing wildfowl do not have to cope with those changes in accessibility of their 'prey' associated with cold conditions in winter. However, the density of food available to them from autumn onwards also decreases, but as a result of their own foraging efforts. How they cope with this, how they choose where to forage, and how annual changes in food stocks from place to place influence the timing of their movements, are discussed by van Eerden in

Chapter 5. Enclosure of the Lauwerszee, an arm of the Waddenzee in the northern Netherlands, provided an opportunity to follow over a 10-year period the effects of provision, and later decline, of a major new food source for wildfowl – the seed stock of *Salicornia* spp.

In the final chapter, Pehrsson summarizes his long-term studies of waterfowl feeding in the estuaries and coastal waters of western Sweden. Although his most detailed information refers to diving ducks feeding upon *Mytilus edulis* (whose distribution and abundance is relatively easy to measure, even subtidally), he shows that it is possible to predict the locations of food stocks and clear water for foraging of herbivorous and fish-eating waterfowl also. Pehrsson argues that populations of some duck species, but not all, are controlled by feeding conditions on the winter quarters.

Although in this section of the book food resources are clearly shown to affect the use of coastal areas by waterfowl, the mechanisms are not treated in detail here, but are explored in Part 2.

1

Coastal birds: numbers in relation to food resources

P. R. EVANS & P. J. DUGAN

Many coastal wetlands and estuaries are threatened by developments. A question often posed by planners and developers is 'if an area is totally changed, so that wildfowl and waders no longer can feed there, can they go elsewhere, or will they die?' Alternatively, if only part of the area is to be developed, the question is likely to be 'can all the birds that normally use the whole site be fitted into the reduced area?' The general answers to both questions are that birds can be accommodated physically in areas smaller than they are seen to use at present and that (if they can find suitable habitats) they can be accommodated elsewhere. However, their chances of survival may well be reduced in either situation, because, as bird density on an area increases, the rate at which each bird can obtain food usually decreases (Goss-Custard 1980).

To provide more detailed answers to these questions we need to ask further questions concerning the present distribution of birds amongst coastal wetlands and estuaries. Most importantly, we need to ask what factors set the limits (if such exist) to the numbers of each species that occur in an area and at what times of year these limits are reached.

In this chapter, we approach these problems by summarizing our present understanding of the relationships between numbers of certain grazing wildfowl and their food supplies. We then consider the more complex interactions between shorebirds and their foods, first by reviewing our current knowledge of factors which affect the ability of waders to meet their energy requirements, and then by examining how these factors may influence the relationships between bird density and food supply. Finally, we consider our present inability to make precise predictions of the effects of reclamation of intertidal areas on numbers of waders using them, and suggest topics on which work is required to rectify this situation.

Coastal grazing wildfowl and their food supplies

In most parts of the northern temperate zone, very little plant or animal growth occurs between late autumn and spring. This means that the food resources of coastal wildfowl and waders during this period are effectively non-renewable and so will be steadily depleted. If vegetation is always accessible to grazers throughout the winter months, as may be true of saltmarsh grasses unless covered with ice or snow, it provides a finite number of bird-days of feeding. If more birds come or are forced to feed on a site, the food stocks will be exhausted more quickly.

For any bird to maintain good nutritional condition, its daily energy and nutrient intake should balance its daily requirements. For this to happen, it must achieve a certain rate of food intake, determined partly by the length of time for which it is able to feed in each 24 hours. For coastal birds, this may not be the full 24 hours because: (i) in intertidal areas food is not available at high water but only during certain parts of each tidal cycle, as in the case of wildfowl grazing sea-grasses (*Zostera* spp.); (ii) a longer time may be needed to digest food than to ingest it, as may happen with geese feeding on the green algae, *Enteromorpha* spp. (Charman 1979); and (iii) feeding may be undertaken safely only at certain hours, e.g. by night, to avoid predation or food-stealing. The minimum acceptable rate of food intake is equal to the daily food requirement divided by the maximum possible feeding time. In general terms, the food intake rate is affected by the density of the food resource; this may be measured more appropriately by biomass (g/m^2) in some instances and by items/m^2 in others. Whichever units are used, there should exist a certain threshold density of leaves below which it is no longer profitable for a bird to feed upon them, because it cannot ingest them at a sufficient rate to achieve its daily requirements, or even balance the cost of foraging.

Threshold densities have been measured in the field for *Zostera* leaves being grazed by Brent Geese (*Branta bernicla*) (Charman 1979) and for mixed grasses being grazed by Barnacle Geese (*Branta leucopsis*) (Ebbinge, Canters & Drent 1975). These have enabled predictions to be made in autumn of the number of goose-days for which *Zostera* and grass meadows, respectively, would be used by the two goose species during the winter according to the formula:

$$\text{Goose-days/unit area} = \frac{\text{vegetation density} - \text{threshold density}}{\text{goose daily requirements}}.$$

These predictions have accorded well with actual measurement of goose-days of use. However, it has not yet been confirmed that the threshold plant density, measured in the field, corresponds to that providing the *minimum* food intake rate required to balance the daily energy budget. Observations on captive geese are required.

When or just before food stocks reach the threshold density, three options are open to the grazing bird: (i) to move to a new feeding area; (ii) to stay but utilize additional types of food; (iii) to stay but utilize a wider size range of the food in scarce supply. Birds that move may, in addition, change their diet. Some Brent Geese that arrive in autumn in Essex, southeast England, move to other areas in southern England when *Zostera* stocks become depleted. Some again feed on *Zostera* in the new sites but others change to a diet of *Enteromorpha* or even graze winter wheat; as do most of those that remain in Essex (Charman & Macey 1978).

Although a considerable amount of information is now available on the factors controlling the number of goose-days that an area of coastal vegetation can provide, little is known of any factors controlling goose numbers feeding on an area at any time. At vegetation densities above the threshold, the average distance between neighbouring geese in a flock, i.e. the flock density, is little affected by food density. Each individual goose in the flock spends less of its time being vigilant for approaching danger (rather than feeding) as flock size increases. However, above a flock size of about 30 birds, the increased feeding time for an individual, resulting from the addition of another bird to the flock, is negligible. Yet flocks are often much larger, and it is not clear what might impose an upper limit to flock size. Geese graze by walking slowly across a feeding area. Their walking speed affects the rate at which individuals encounter food items. Those at the back of the flock must utilize areas already partially grazed a few minutes earlier by birds at the front of the flock. The 'depth' of the flock may thus be limited by the need for the density of vegetation available to birds at the back to be just above the threshold, but it is not clear whether the width of the flock is limited.

Thus the question 'how many geese can an area support' rather than 'how many goose-days can a feeding area provide' has yet to be answered. Possibly it is not a biologically meaningful question, because there may be no mechanism limiting flock size.

So far, it has been assumed that different species of grazing wildfowl in coastal areas do not compete for the same plant food. This, however, is not always the case. For example, at Lindisfarne, Northumberland, *Zostera* spp. are grazed not only by Brent Geese but also by Wigeon

(*Anas penelope*) and Whooper Swans (*Cygnus cygnus*). There is slight segregation of feeding areas (Fig. 1.1), but on most areas indirect competition by exploitation occurs, with different species feeding on the same resources at different times of the tidal cycle. Clearly, the daily food requirements of a swan weighing 9 kg are considerably greater than those of a goose of 1.3 kg which in turn are greater than those of a Wigeon of 0.5 kg. When this is allied to differences in feeding method (affecting the rate of food intake) and differences in potential feeding time, one would expect differences in the threshold density of *Zostera* at which each species could exploit the sea-grasses profitably.

In practice, most of the 30 000 Wigeon which spend the autumn at Lindisfarne leave the mudflats and move further south in England before most of the geese and swans arrive, though whether they leave because the density of *Zostera* on the beds they have exploited has decreased to below the threshold density they can utilize is not known. A further complication, also illustrated by the situation at Lindisfarne,

Fig. 1.1. Feeding sites of Wigeon (*Anas penelope*), Brent Geese (*Branta bernicla*) and Whooper Swans (*Cygnus cygnus*) at Lindisfarne, northeast England (redrawn from Smith 1977).

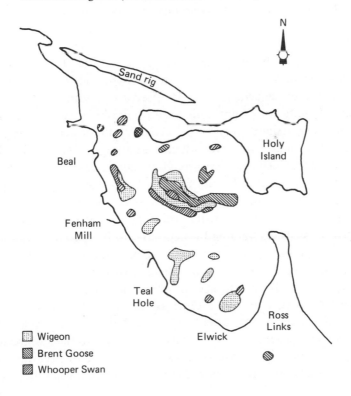

is that the reduction in food stocks available to herbivores may be due not only to the grazers themselves but also to natural die-back of vegetation. Of the two species of *Zostera* at Lindisfarne, only *Z. nana* retains its leaves throughout the winter. The leaves of *Z. marina*, though available to wildfowl in October, die back quickly thereafter, and only roots are then available. These are utilized by Whooper Swans but not by the ducks or geese. Therefore, prediction in the autumn, before the Wigeon arrive, of the number of Brent Goose-days or Whooper Swan-days that can be provided later in the winter is clearly unreliable in our present state of knowledge.

Wildfowl may take not only the leaves and roots of plants, but also their seeds, by a variety of feeding techniques. Some of these are considered in Chapter 5 of this volume, in which van Eerden concludes that no mechanism appears to exist to restrict the *numbers* of wildfowl using a single seed resource (that of *Salicornia* spp.), although the number of bird-days of use are limited. When several species use the same resource, the relative numbers of bird-days of use by each depends in part on their dates of arrival in autumn.

Coastal shorebirds and their food supplies: some general considerations

In most coastal habitats, plant foods are always accessible to grazers, except if immersed periodically by the sea. In contrast, the invertebrate foods of waders, although present, may not always be accessible to, or detectable by, the predators. Some waders – for example, the plovers – hunt their prey primarily by sight. They have short bills and require prey to be active at the surface of exposed intertidal sands and muds, over which they forage. Other waders hunt chiefly by touch, aided sometimes by visual detection of where to probe. These species – such as Dunlin and Curlew – have longer bills and obtain their prey from below the surface of the mud, though they require prey to lie at depths within the range of their bills before they can catch them. Birds hunting by sight tend to be spaced further apart than those hunting by touch, thus reducing interference from others while catching prey (Goss-Custard 1970*b*). Interference increases the time for which birds must forage to obtain their food requirements.

In soft, exposed intertidal sediments, the depths at which invertebrates lie and their activity often vary throughout the tidal cycle. Many worms and crustaceans are most active and closest to the surface when the tide just uncovers or covers the sediments. For this reason, many species of waders that feed upon them forage by following the edge of the ebbing and flowing tide. What is more important, however, is that as the temperature falls many invertebrates, including some bivalves, dig deeper and, in particular, their activity decreases. This alters the percentage of prey available to a predator during the winter in the way shown in Fig. 1.2. (For a given prey, the position of the 'percentage available' line varies from one species of predator to another, depending on the hunting method.)

In most northern temperate coastal zones, invertebrates neither grow nor reproduce from autumn to spring, so that their absolute density falls somewhat through predation and death during this period. However, this effect is overridden by the changes in their availability associated with temperature, so that the density of *available* prey usually sinks to a minimum when it is coldest. This is normally in mid-winter.

Mud- and sand-flat invertebrates do not occur at uniform density throughout an estuary but are concentrated into patches, often of much smaller area than those of coastal vegetation. Shorebirds, like grazing wildfowl, must achieve a certain minimum rate of prey capture to meet their daily energy requirements. This they can do only by feeding in areas of a sufficiently high density of available prey (Fig. 1.3).

However, prey availability, and hence the size of the area in which feeding is profitable, fluctuates from day to day in association with changing weather conditions. Furthermore, in mid-winter when it is coldest, not only is the percentage availability of prey lowest, when

Fig. 1.2. Changes in absolute density (—) of invertebrate prey, in percentage availability (····) and hence in density of available prey (---) during an average winter.

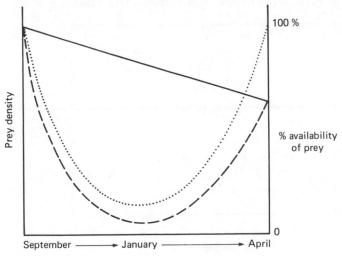

averaged over a day, but the minimum acceptable rate of prey capture is highest, because the daily costs to the bird of maintaining an elevated and constant body temperature are highest then. Thus the total area of habitat in which the density of available prey is sufficiently high for profitable feeding will be much less in mid-winter than in the autumn or spring. Under cold conditions, two things might happen: (i) birds might maintain the same spacing as in mild weather, so that only those individuals established on feeding areas with the highest prey densities could feed profitably; or (ii) birds might tolerate closer nearest neighbours, so that the density of birds in areas of higher prey density would increase. This second alternative, however, would be likely to lead to a reduction in the rate at which any individual could capture food, either because prey became less active in response to increased predator activity on the mud surface, or because predators interfered with each other's ability to capture prey (Goss-Custard 1980). (This assumes that, during mild weather, bird density in areas of high prey density is not regulated well below the level at which interference occurs, for example by territorial behaviour.) Pienkowski (1980) found that as temperatures fell at Lindisfarne, Grey Plovers (*Pluvialis squatarola*) caught prey less frequently. In cold weather they gathered on a few areas of high prey density but, within these areas, maintained larger distances between individual birds as temperatures became progressively lower.

Temperature is not the only meteorological factor to affect a shorebird's chances of maintaining energy balance. Stronger winds not

Fig. 1.3. The influence of the density of available prey on the rate of prey capture by Oystercatchers (redrawn from Hulscher 1976).

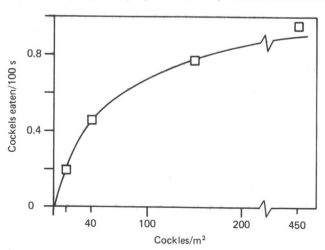

only increase the rate of heat loss from a bird, and hence its daily energy requirements, but may prevent a bird from locating cues given by certain prey at the mud surface and so prevent feeding altogether (Dugan *et al.* 1981).

It is clear from this discussion that the density of shorebirds that could in theory be supported by a certain absolute density of prey varies from place to place in relation to the seasonal changes in temperature and wind strength experienced at each, not only through the effects of temperature and wind strength on energy demands of the birds, but also on prey behaviour. These latter effects will vary from estuary to estuary according to the prey species present and their various responses to changes in temperature. Also, the availability of a particular prey species may differ amongst different substrate types.

The seasonal storage by shorebirds of nutritional reserves (see Chapter 2 of this volume), as an adaptation which enhances survival during short periods of severe weather, introduces further complexity into the relationships between the birds and their prey. If birds are unable to feed profitably, or to obtain an adequate intake in the short term, they can draw upon their fat reserves, which will provide energy sufficient to enable survival during a few days of even total starvation, resulting from temporary unavailability of prey. These reserves are regulated at different levels in different months. When laying down fat in autumn, birds must achieve a higher daily food intake than that required to maintain body weight, but in spring can manage with a lower rate since their fat reserves are being run down. Similarly, higher rates are required immediately after a period of severe weather, to replace depleted reserves. To enable deposition or re-deposition of reserves, birds need to feed in areas of higher densities of available prey and for longer times than are required simply to maintain body weight. The implication of this is that any attempt to relate bird density to average prey density in different estuaries that takes account only of the more obvious seasonal effects (e.g. of average temperatures) on prey availability and energy requirements of the predator must remain superficial.

Since most waders (unlike some grazing wildfowl) do not swim whilst feeding, their foraging is confined to the exposed intertidal flats and to shallow water at the tide edge. This means that the total foraging area available to them changes markedly from a large expanse at low water to a narrow strip at high water. Added to this, the densities of many benthic invertebrates decrease towards the high water mark. Not only do birds require a minimum density of available prey to permit profitable feeding, but also they are unable to increase their rate of food intake as prey density increases above a certain plateau level (Fig. 1.3), as found, for example, by Hulscher (1976).

Probably, this plateau rate of food intake is achieved only during milder weather conditions. In cold and windy weather, as their energy requirements increase, birds usually have to extend their feeding hours, rather than feed faster. (Indeed, their rate of food intake is liable to fall as prey availability decreases.) Thus, under the most difficult feeding conditions, they need to use not only their regular feeding sites but also those at higher tidal levels than usual, where patches of prey of available densities above the threshold are scarce. Hence the crucial determinant of relationships between wader numbers and food resources in severe weather may well be the extent of high-density patches of available prey at high tidal levels. If feeding grounds at lower tidal levels are covered early on the flood tide, they may not provide enough feeding *time* for the birds to survive in mid-winter (see the discussion by Swennen in Chapter 10 of this book). The length of tide edge at high tidal levels may also affect feeding performance, by affecting the degree of crowding of feeding birds (Fig. 1.4). At the tide edge, activity and accessibility of many prey species is highest, so that birds feeding upon them concentrate there (Evans 1979). The longer the tide edge, the greater the spacing of birds and the better the feeding rate achievable by each. (Exceptions to this are provided by shorebirds such as Oystercatchers (*Haematopus ostralegus*) that feed on sedentary bivalves, e.g. the cockle *Cardium edule*, whose availability is not necessarily highest at the tide edge.)

In addition to these effects of estuary shape, tidal amplitude may greatly influence the numbers of certain shorebirds that can survive the winter in a particular site. In some areas, e.g. the Waddenzee, where

Fig. 1.4. The effects of differences in beach profile on the length of tide edge at different tidal levels.

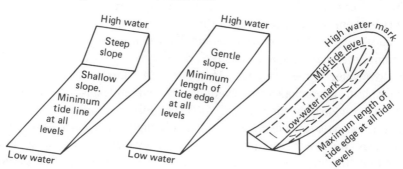

the amplitude is very small (only 1–2 m between low and high water), on-shore winds may hold back the receding tide. In extreme cases when this happens, the feeding grounds of many short-legged shorebirds remain completely covered. If such conditions coincide with low temperatures, they could lead to rapid depletion of a bird's nutritional reserves and perhaps its death. Thus, despite high densities of available prey in these areas when they are exposed at low water, birds may not overwinter there, but instead may favour areas where exposure times of the feeding grounds are more predictable, even if densities of prey there are lower.

Thus in attempting to relate shorebird numbers to prey density in different areas, yet another set of factors – the physical characteristics of the areas – may confound any relationships that would otherwise be demonstrable. Often related to these physical characteristics is the problem of human disturbance. Waders and coastal wildfowl are birds of open habitats. Building of dykes, reclamation walls and flood defences affects the visibility afforded to ground-feeding birds. Some species, particularly the heavier ones which require slightly more time to become airborne, e.g. the Shelduck *Tadorna tadorna*, will not normally feed close to high walls or in enclosed areas. Consequently, this may preclude their use of narrow estuaries, particularly if these are bordered by flood protection barriers.

The picture built up so far has assumed that, like vegetation, invertebrate food resources are sedentary. In the case of some polychaetes and many bivalves living deep in soft sediments (e.g. clams (*Mya arenaria*) – see Chapter 4 of this volume), this is a reasonable assumption. It is also probably true that most prey species are least active and least available in mid-winter, though not enough field studies have yet been made to confirm this. More importantly the behaviour, and therefore availability, of benthic intertidal invertebrates at night has not been studied. Most wader species are able to feed by night, but the relative contributions of day and night feeding to the total daily energy requirements have not been assessed accurately in the field except in the case of Grey Plovers and Ringed Plovers (*Charadrius hiaticula*), in which night feeding may be at least as important as day feeding in mid-winter (Pienkowski 1982; Dugan 1981*b*; see also Chapter 2 of this volume). The limited information available so far indicates that some individuals may feed at different sites on an estuary and on different prey during the night from those used during the day. This enhances the difficulties in attempting to relate numbers of waders on an estuary to food resources.

All the information presented hitherto has been discussed on the assumption that the densities of one bird species do not affect the densities of another, provided that they do not compete for the same prey. In practice, diets of several wader species may overlap, as also described earlier for wildfowl at Lindisfarne. Even if this occurs, partial segregation of feeding areas may occur, other than by direct competition for space. For example, the same absolute density of prey provides different densities of available prey, depending on the foraging method of the predator, visual or tactile, and the length of its bill. In addition, some wader species may be able to exploit, profitably, lower densities of available prey than others, and hence exploit larger areas within each patch of prey. Small species need less food per day than large ones; and species that forage by stand-and-wait tactics may need to expend less energy in foraging than active searchers, and again need less food per day. Thus small, inexpensive foragers might feed profitably in areas of lower density of available prey than large active foragers.

In spite of these mechanisms, whereby direct competition between species for space may be restricted only to those areas holding the highest densities of available prey (which should be the preferred feeding areas), other factors may be important. Although no evidence has yet been produced of interspecific territoriality in waders (e.g. at Teesmouth, Curlew (*Numenius arquata*) and Grey Plover territories may overlap, although both are feeding on *Nereis*), physical crowding by one species may restrict the size of feeding areas used by another. For example, when a wreck of sprats (*Clupea sprattus*) occurred in February 1977, many thousands of gulls (*Larus* spp.) moved on to the sea beaches at Teesmouth and prevented Sanderling (*Calidris alba*) from feeding there, even though the Sanderling were taking small and (for gulls) unprofitable prey. Similar interactions leading to segregation of feeding areas used by flocks of Avocets (*Recurvirostra avocetta*) and Black-headed Gulls (*Larus ridibundus*) have been reported by Zwarts (1978). Dunlin (*Calidris alpina*) flocks have been observed to displace solitarily feeding Redshank (*Tringa totanus*) from mudflat sites at Teesmouth (Moumoutzi 1977).

Seasonal changes in shorebird numbers in relation to prey abundance

Having discussed the many and varied situations faced by shorebirds whilst attempting to maintain energy balance at different times of year, we can now examine the effect of these on the relationship between bird density and prey density by considering one crucial question: since prey availability decreases from autumn to winter, and at the same time predator energy requirements rise, does the density of each wader species found on a feeding site fall steadily in parallel with the fall in density of available food? Detailed studies on the Tees estuary have established seasonal patterns of numbers for seven species of waders and for the Shelduck (Evans 1981*a*). Only in the case of Redshank, and perhaps of Curlew, do bird numbers follow the predicted changes in food availability. This leads directly to the conclusion that for most species the estuary is not filled to its 'potential carrying capacity' in the autumn, i.e. more birds could be supported at this time of year than are found on the estuary. This pattern could arise if not enough birds reach the potential wintering site until later in the year because it is at the 'end' of a migration route. However, while this may be true of Shelduck, there is ample evidence of passage of, e.g., Dunlin and Grey Plovers through the estuary in autumn, so that at least some birds are 'programmed not to stay' or are prevented from doing so.

In the case of Dunlin at Teesmouth, birds from Iceland pass through in August and early September, to be replaced then by birds from western Siberia, at first chiefly juveniles, but from October onwards, adults also (Goodyer & Evans 1980). Similarly, some Grey Plovers arriving on the Tees in August and September move on rapidly. In contrast, other individuals do not leave until later in the autumn, after they have held territories for several weeks. These may be individuals that are ineffective in preventing neighbouring territory holders from foraging in their territories. Later in the winter, however, when more Grey Plovers arrive, the number of birds defending territories increases still further, despite the reduction in prey density and availability. To account for this Dugan (1981*a*) suggested that, to provide the maximum safeguard against severe conditions, a territorial plover defends an area as large as it can, and whenever possible will intrude into a neighbour's territory. The size of each territory is presumed to be limited by the individual's competitive ability and the number of other birds attempting to establish territories. The late winter increase in number of birds holding territories is thus attributed to an influx of birds, including some of at least equal competitive ability to those territorial individuals already present.

Thus, in summary, although the seasonal changes in numbers of a few shorebird species follow the changes in density of available prey, this is not the case for most species. In some of the latter species, the pattern of change in bird numbers is dependent upon external factors which lead to movement of birds from other areas, and in some species may be determined by the social behaviour of birds on the estuary. In view of this, we must change our question to ask whether, when birds are present, they concentrate where prey density is highest.

The relationship between prey density and bird density

In view of the complex array of factors which can affect the chances of an individual shorebird achieving energy balance by feeding on a particular site, it is not surprising that the most convincing relationships between wader densities and the densities of their food resources have been demonstrated on a small geographical scale, i.e. within an estuary. For example, O'Connor & Brown (1977) related numbers of Oystercatchers feeding on different parts of Strangford Lough, Northern Ireland, to the densities of cockles, their only prey there. In a more detailed study, Goss-Custard (1970a) examined the densities of Redshank feeding on different sections of the Ythan estuary, northeast Scotland, in relation to the absolute densities of their chief prey, the amphipod *Corophium volutator*. At low prey densities, Redshank densities were proportional to *C. volutator* densities; however, at high prey densities, Redshank densities reached a plateau level. This may be generally true if behavioural interactions between birds limit the rate at which each individual can obtain food (at bird densities above a certain level), or if only part of the population on an estuary shows territorial behaviour – perhaps only on certain types of substratum (see Chapters 3 and 8 of this volume for further examples).

On a wider geographical scale, Wolff (1969) related the distributions of different wader species in the estuaries of the Dutch Delta area to the distributions of their invertebrate prey. (Different bird species tended to take different prey.) Associated with differences in salinity between different sites, there was a considerable degree of segregation of the more abundant invertebrate species, and corresponding segregation of their bird predators. Unfortunately, on an even wider geographical scale, the prey species selected by a given species of wader may vary from place to place. This arises because the species composition of the intertidal invertebrate community often varies from country to country, and because each wader takes those of the prey species present in a feeding area that are most available to its particular feeding method (Pienkowski 1981). Furthermore, the diet of each wader species may change from year to year in a single locality, if there are marked annual changes in density of different prey. Hence the Dutch results, though an entirely valid description of the situation in the Delta area in the year of study, cannot be generalized safely.

More recently, Goss-Custard, Kay & Blindell (1977) examined the quantitative distribution of waders, particularly Curlew and Redshank, between nine adjacent areas of the coasts of Essex and Suffolk, extending over about 50 km. They found that the average winter densities (between November and March) of the two bird species were directly proportional to the average densities of their main prey in these nine sites. The high values of the coefficients of correlation between absolute densities of predator and prey (about 0.7 for Redshank and 0.9 for Curlew) are rather surprising, since the invertebrate densities in different estuaries were measured at different times of year and, for Redshank, the seasonal changes in numbers of birds did not parallel the expected changes in food availability. Furthermore, the birds were counted at roosts and not on the intertidal feeding areas; in the mild and wet winter of 1973–74, to which the data refer, it is likely that many birds fed in mid-winter in coastal fields and brackish marshes, although roosting with the intertidal feeders, as happened on the Tees estuary (Evans 1981*a*).

The associations demonstrated by Goss-Custard *et al.* (1977) were simple correlations between absolute densities of birds and absolute densities of their main prey. By summing for each site the absolute densities of the two or three important prey taken by each bird species, the correlations with bird densities were improved. They concluded that the correlations had a low probability of being due to chance, and that study of the possible processes by which the numbers of birds in the estuary could become related to the density of food were needed. Some recent studies on waders are reported elsewhere in this volume (e.g. Chapters 10 and 11).

It is possible that initial selection of potential feeding sites is achieved not from the ground, but from the air. We speculate that Curlew arriving in southeast England from Scandinavia might well fly at an altitude of 1.5 km, which would place their visual horizon at a range of about 140 km, so that they should be able to assess the visual characteristics of coastal sites (e.g. extent of mudflats, sandflats, exposure to wave action, etc.) over considerable distances. If indeed the birds have selected their arrival sites in this way, it would explain why the Suffolk and Essex estuaries can be treated successfully as a single unit for study of bird distribution amongst them.

In an attempt to determine whether the conclusions of Goss-Custard *et al.* (1977) could be extrapolated to other areas, Bryant (1979) examined the distributions of six species of waders between 14 adjacent sites along both banks of about 32 km of the inner Firth of Forth estuary in southeast Scotland. From counts of feeding birds, he was able to relate both bird densities and bird feeding-hours/km² to invertebrate densities at each site. As in southeast England, he found significantly higher densities of birds in those sites holding higher densities of certain invertebrates (but his correlations were based on double logarithmic scales, which do not distinguish between straight-line relationships of the type found by Goss-Custard *et al.* (1977) and relationships in which bird densities reach a plateau, as found by Goss-Custard (1970*a*)). Some of Bryant's simple correlations, for example between Redshank density and *Corophium* density, disappeared when bird distributions were examined in relation to the densities of several invertebrate species and several site characteristics (e.g. area, disturbance index) simultaneously, in a multiple regression analysis. Conversely, a few other relationships, for example between Redshank density and *Hydrobia* density, which had not been revealed by simple correlation analysis, became significant in the more complex analysis.

The two studies described above were still on a relatively small geographical scale. Within a localized area, such as a single, albeit large, estuary like the Forth, or a complex of small estuaries like those in Essex and Suffolk, where tidal and weather conditions are relatively uniform, the overriding importance of prey density as a predictor of predator density is not surprising. However, because differences in tidal and weather conditions are of much greater importance on a wider geographical scale, the value of such correlation analyses as a means of explaining distribution on a larger scale, e.g. within the whole of the British Isles, seems doubtful. Nevertheless, by restricting comparisons to the simple correlations between Redshank density (or Curlew density) and density of the ragworm *Nereis diversicolor*, Bryant suggested that the relationships found in the Forth and in southeast England were sufficiently similar to indicate that this type of analysis might be used to predict bird densities at other sites, given the relevant invertebrate densities (without considering the problem of availability). However, although the slopes of the trend lines relating Curlew densities to *Nereis* densities appear similar in the two areas, the data had been plotted on double logarithmic scales, so that the small difference in intercepts (Fig. 1.5) indicates that Curlew occur at densities some 10 times higher in Suffolk and Essex than in Scotland. The direction of this difference is what might be expected on theoretical grounds but the magnitude is not. The northern limit of a species' winter distribution in the northern hemisphere must be influenced in part by weather factors, reducing the percentage availability of prey and increasing the energy requirements of the predators. However, the minimum January temperatures are only 1 °C lower on average on the Forth than in Essex and Suffolk (Dugan 1981*a*). In view of this, the tenfold difference in density of Curlew between the two areas is surprisingly large. Before useful extrapolation can be made to other areas, a much fuller understanding is required of the processes underlying the observed relationships.

We have indicated that in some localized areas a relationship can be demonstrated between bird density and absolute densities of prey. However, the precise form of this relationship must vary with latitude because of differences in average temperatures and other weather factors, and hence in prey availability and energy requirements of the birds. Furthermore, for species showing some degree of territorial behaviour, the form of the relationship may depend on the distribution of competitive abilities within each wintering area and on the proportion of birds defending feeding sites in each area. However, demonstration of such relationships does not help to resolve the question 'is the estuary full: would addition of more birds reduce the chances of survival of those already present?' This question may be illustrated in the following way (Fig. 1.6).

Do the two lines coincide, or is bird density held below the maximum level, and if so, by what mechanisms? As we have described earlier, Grey Plover density on the Tees estuary in autumn is held below the

Fig. 1.5. The relationship between Curlew density and density of its prey, *Nereis diversicolor*, on estuaries in southeast England (triangles) and southeast Scotland (circles) (redrawn from Bryant 1979).

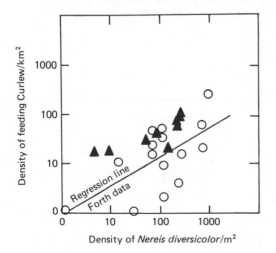

maximum level in those parts of the estuary where territories are established. However, we cannot yet predict to what extent additional birds, attempting to establish territories in late winter, reduce the chances of survival of territorial birds already present. It is probable that, by failing to obtain a territory at this time, an individual's chances of survival are reduced, since non-territorial birds find it more difficult than territorial birds to maintain energy balance in severe weather conditions in late winter (Dugan 1981*a*). Thus, if more birds than usual come to the Tees estuary from other estuaries in late winter (for whatever reason), or if the extent of the feeding area at Teesmouth is further reduced by reclamation, in both cases a smaller *proportion* of birds present would manage to obtain territories, and therefore a higher proportion of birds would be non-territorial, and at risk from severe weather. This would lead to a greater reduction in overall numbers of birds on the estuary after a severe winter than would have occurred before reclamation took place.

So far, only one shorebird species (the Oystercatcher – see Chapter 3 of this volume) has been proved to respond to its food resources in the relatively simple way described for Teal (*Anas crecca*) by Zwarts (1976). He noted that birds settled on an estuary in autumn by filling first the more profitable feeding areas. Once a ceiling density of feeding birds had been reached, other arrivals were unable to feed in these sites, although allowed to stay nearby; they did not, however, remain for long. Occupation of feeding areas in sequence has also been shown for

Fig. 1.6. Hypothetical relationships between shorebird density and prey density.

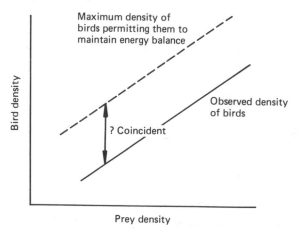

Knot (*Calidris canutus*) and Oystercatchers on the Wash (Goss-Custard 1977), but the relative profitability of the areas was not known.

Further complications: responses of shorebirds to mobile prey

Although the availability of sedentary invertebrates to birds may change seasonally and from day to day in response to changes in temperature and other weather factors, shorebirds feeding on them could, in theory, predict the changes that are likely to occur in an average winter (see Fig. 1.2) and allow for them when they settle on an area in autumn. For example, they might assess the quality of a potential feeding area when they first arrive and stay there only if the density of available prey in autumn is, e.g., 100 times the threshold density needed to provide an adequate rate of food intake in mid-winter. No such allowance can be made if the change in density or availability of food is totally unpredictable. This is most likely to occur in invertebrate populations that inhabit unstable, soft sediments, notably on exposed sandy beaches. Here the beach profiles may change dramatically overnight as wave and wind action move large quantities of sand and the associated invertebrates. Their bird predators can insure against such happenings only by their own mobility. For example, some colour-marked Sanderlings have been found to patrol many tens of miles of coastline in northeast England, whereas others remain sedentary at the Tees estuary. It has been suggested that the patrolling birds are continually checking the quality of several potential feeding areas to guard against the sudden and unpredictable removal of beach sediments at Teesmouth by wind and tide, and the associated removal of food resources from that site. Only rarely, perhaps not every year, does this occur on a sufficiently large scale to lead to food shortage for the birds that stay at Teesmouth (Evans 1981*b*). Other species also show mobility during the winter months, leaving one estuary before available food resources decrease to levels at which mortality occurs, e.g. Knot in eastern Britain (Dugan 1981*a*).

Unpredictability of food resources may occur in space as well as in time. Many intertidal crustaceans move about during the high-water period, when they are covered by the tide, and although they may settle on the retreating tide at the same level downshore, concentrations may shift alongshore, influenced by water currents and wind action. Detailed studies of the distribution of Sanderling feeding on the amphipod *Excirolana* on coastal beaches in California revealed that some birds defended territories whilst others fed non-territorially. To guard against the effects of movement of their prey, the territorial birds defended areas as large as possible, even though these often contained much more food than they required (Myers, Connors & Pitelka 1979).

Thus, if the prey are mobile, or are liable to be removed at unpredictable times by physical forces, the strategies adopted by shorebirds to enhance their own survival are likely to lead to a total absence of reproducible correlations between bird density and prey density measured over more than short time periods. In the case of Sanderling patrolling a coastline, during any period in which sediment (and therefore invertebrate) movement is minimal, the type of correlation most likely to be found might well be between the duration of stay at a feeding locality and the density of prey there; but over longer time periods no correlations may be demonstrable.

Conclusions: the direction of further work

It will be clear from this review that the greatest gaps in our knowledge of the relationships between coastal birds and their food resources lie in our ability to predict the densities of shorebirds likely to be found in a site holding a given density of prey. Because the factors affecting prey availability and shorebird survival are so many and complex, simple models of predator–prey density relationships are not likely to give accurate predictions. This situation can be improved in two ways: (i) by empirical studies of the relationships in many different estuaries, so that factors such as differences in prey taken, temperature, tidal amplitude, wind conditions, substrate type, etc. can be built into a more sophisticated predictive model; and (ii) by detailed studies of the mechanisms by which the density of a particular wader species is adjusted to the density of its prey.

We have suggested that changes in the density of birds using a given density of prey are likely to lead to changes in survival rate amongst the birds. It will be important to establish which types of individuals are most at risk – males?, females?, juveniles? or perhaps those of a particular size in relation to the average size? Such information may be obtainable when bird densities are increased naturally during severe weather, e.g. as a result of restriction of feeding areas to those free from ice-cover. It will also be necessary to have this information if predictions of the long-term effects of reduction in intertidal land are to be reliable.

Finally we need to remember that several shorebird species are very mobile during the non-breeding season so that, in these at least, predator–prey density relationships are likely to vary widely from time to time, as also they may in species whose populations on one estuary include varying proportions of individuals showing territorial behaviour at different times of year.

Acknowledgements
We thank the Natural Environment Research Council, the Nature Conservancy Council and the Nuffield Foundation for financial support for our studies at Teesmouth, and our past and present colleagues, particularly Drs M. W. Pienkowski, N. C. Davidson and D. J. Townshend for comments on earlier drafts of this review.

References

Bryant, D. M. (1979). Effects of prey density and site character on estuary usage by overwintering waders (Charadrii). *Est. coast. mar. Sci.* **9**: 369–84.

Charman, K. (1979). Feeding ecology and energetics of the Dark-bellied Brent Goose (*Branta bernicla bernicla*) in Essex and Kent. In *Ecological processes in coastal environments*, ed. R.L. Jeffries & A. J. Davy, pp. 451–65. Blackwell, Oxford.

Charman, K. & Macey, A. (1978). The winter grazing of saltmarsh vegetation by Dark-bellied Brent Geese. *Wildfowl* **29**: 153–62.

Dugan, P. J. (1981*a*). Seasonal movements of shorebirds in relation to spacing behaviour and prey availability. Unpublished PhD thesis, University of Durham, South Road, Durham DH1 3LE, England.

Dugan, P. J. (1981*b*). The importance of nocturnal foraging in shorebirds: a consequence of increased invertebrate prey activity. In *Feeding and survival strategies of estuarine organisms*, ed. N. V. Jones & W. J. Wolff, pp. 251–60. Plenum Press, New York.

Dugan, P. J., Evans, P. R., Goodyer, L. R. & Davidson, N. C. (1981). Winter fat reserves in shorebirds: disturbance of regulated levels by severe weather conditions. *Ibis* **123**: 359–63.

Ebbinge, B., Canters, K. & Drent, R. H. (1975). Foraging routines and estimated daily food intake in Barnacle Geese wintering in the northern Netherlands. *Wildfowl* **25**: 5–19.

Evans, P. R. (1979). Adaptions shown by foraging shorebirds to cyclical variations in the activity and availability of their invertebrate prey. In *Cyclic phenomena in marine plants and animals*, ed. E. Naylor & R. G. Hartnoll, pp. 357–66. Pergamon Press, Oxford.

Evans, P. R. (1981*a*). Reclamation of intertidal land: some effects on Shelduck and wader populations in the Tees estuary. *Verh. orn. Ges. Bayern* **23**: 147–68.

Evans, P. R. (1981*b*). Migration and dispersal of shorebirds as a survival strategy. In *Feeding and survival strategies of estuarine organisms*, ed. N. V. Jones & W. J. Wolff, pp. 275–90. Plenum Press, New York.

Goodyer, L. R. & Evans, P. R. (1980). Movements of shorebirds into and through the Tees estuary, as revealed by ringing. *Cleveland County Bird Rep.* **6**: 45–52.

Goss-Custard, J. D. (1970*a*). The responses of Redshank *Tringa totanus* to spatial variations in the density of their prey. *J. anim. Ecol.* **39**: 91–114.

Goss-Custard, J. D. (1970*b*). Feeding dispersion in some overwintering wading birds. In *Social behaviour in birds and mammals*, ed. J. H. Crook, pp. 1–35. Academic Press, London.

Goss-Custard, J. D. (1977). The ecology of the Wash III. Density-related behaviour and the possible effects of a loss of feeding grounds on wading birds (Charadrii). *J. appl. Ecol.* **14**: 721–39.

Goss-Custard, J. D. (1980). Competition for food and interference among waders. *Ardea* **68**: 31–52.

Goss-Custard, J. D., Kay, D. G. & Blindell, R. M. (1977). The density of migratory and overwintering Redshank *Tringa totanus* and Curlew *Numenius arquata* in relation to the density of their prey in south-east England. *Est. coast. mar. Sci.* **5:** 497–510.

Hulscher, J. B. (1976). Localization of cockles *Cardium edule* by the Oystercatcher *Haematopus ostralegus* in darkness and daylight. *Ardea* **64:** 292–311.

Moumoutzi, L. (1977). A study of the feeding distribution of Dunlin on Seal Sands during the spring and late summer. Unpublished MSc thesis, University of Durham, South Road, Durham DH1 3LE, England.

Myers, J. P., Connors, P. G. & Pitelka, F. A. (1979). Territory size in wintering Sanderlings: the effects of prey abundance and intruder density. *Auk* **96:** 551–61.

O'Connor, R. J. & Brown, R. A. (1977). Prey depletion and foraging strategy in the Oystercatcher *Haematopus ostralegus*. *Oecologia* **27:** 75–92.

Pienkowski, M. W. (1980). Aspects of the ecology and behaviour of Ringed and Grey Plovers, *Charadrius hiaticula* and *Pluvialis squatarola*. Unpublished PhD thesis, University of Durham, South Road, Durham DH1 3LE, England.

Pienkowski, M. W. (1981). Differences in habitat requirements and distribution patterns of plovers and sandpipers, as investigated by studies of feeding behaviour. *Verh. orn. Ges. Bayern* **23:** 105–24.

Pienkowski, M. W. (1982). Diet and energy intake of Grey and Ringed Plovers in the non-breeding season. *J. Zool., Lond.* **197:** 511–49.

Smith, P. C. (1977). Main feeding locations of herbivorous wildfowl at Lindisfarne. In *Ecology of Maplin Sands*, ed. L. A. Boorman & D. S. Ranwell, p. 47. Institute of Terrestrial Ecology, Cambridge, England.

Wolff, W. J. (1969). Distribution of non-breeding waders in an estuarine area in relation to the distribution of their food organisms. *Ardea* **57:** 1–28.

Zwarts, L. (1976). Density-related processes in feeding dispersion and feeding activity of Teal *Anas crecca*. *Ardea* **64:** 192–209.

Zwarts, L. (1978). Intra- and inter-specific competition for space in estuarine bird species in a one-prey situation. *Proc. 17th Int. Orn. Congr. (Berlin)* 1045–50.

2

Balancing the budget: measuring the energy intake and requirements of shorebirds in the field

M. W. PIENKOWSKI, P. N. FERNS,
N. C. DAVIDSON & D. H. WORRALL

Introduction

For most of the year, adequate supplies of nutrients are probably obtained by wading birds simply by fulfilling their daily requirements for energy, because they feed mainly on animal food with a high protein content. However, during certain periods some dietary components may not be present in adequate quantities in ordinary foods. For example, calcium may be in short supply when females are forming egg-shells. This increased demand for calcium may be fulfilled by ingesting egg-shells or lemming-bones (Maclean 1974; Byrkjedal 1975). In general, however, shorebird survival and reproduction depends on achieving a satisfactory energy balance.

This chapter is concerned mainly with problems in estimating the energy intake and expenditure by shorebirds in the field, and in considering over what periods of time the balance of intake and requirements needs to be assessed. This additional complication arises because birds store energy (mainly in the form of fat) at certain times during the year and use these stores at other times.

We will illustrate these problems mainly with examples drawn from studies of Ringed Plovers (*Charadrius hiaticula*) and Grey Plovers (*Pluvialis squatarola*) at Lindisfarne, Northumberland, northeast England (Pienkowski 1980a, 1981a,b, 1982, 1983), and of Dunlins (*Calidris alpina*) in the Severn Estuary, South Wales (Worrall 1981). These species are representative of the two main groups of shore waders, the visual-foraging plovers (Charadriidae) and the sandpipers (Scolopacidae). Most of the latter include a large tactile element in their search for prey. Discussions of energy reserves are based mainly on the work of Davidson (1981a,b, 1982).

Energy requirements

The amount of energy utilized by a fasting animal at rest in a thermoneutral environment is usually referred to as the basal metabolic rate (BMR). (The zone of thermoneutrality is that range of ambient temperatures in which body temperature is maintained by adjusting the effectiveness of the body insulation, rather than by metabolic activity.) At lower temperatures (i.e. below about 20 °C in birds), body temperature is maintained within its normal limits by increased metabolic activity.

The BMR and the further energy used for temperature maintenance and cage-living (using food supplied, rather than searched for) may be measured fairly readily with captive birds. Within taxonomically similar groups of species, the BMR has been found to bear a close relationship to body weight (e.g. Lasiewski & Dawson 1967; Kendeigh, Dol'nik & Gavrilov 1977). If the wader species with which we are concerned have a different relationship between body weight and BMR to that of most other non-passerines, our arguments will need some amendment. This is currently the subject of investigation (A. G. Wood, unpublished).

The energy requirements of free-living birds exceed those of caged ones because of the energetic costs of locating, capturing and consuming natural prey, flying to and from roosts and escaping from or avoiding predators, as well as of social interaction (including territorial behaviour), migrating, moulting and breeding. The daily energy budget (DEB), also known as the average daily metabolic rate, includes these requirements, as well as those mentioned earlier, and is thus the most significant measure of metabolism from the viewpoint of an avian field ecologist.

The DEBs are extremely difficult to measure directly, though estimates have been obtained for some passerines by the D_2O^{18} method (Utter & Lefebvre 1973; Hails & Bryant 1979; Bryant & Westerterp 1980), which involves administration of 'labelled' heavy water to individuals and their subsequent recapture after 24 or 48 hours of activity. Recapture of D_2O^{18}-labelled waders has proved to be difficult. The method is also expensive and time-consuming. Telemetry of heart rate, using a small radio transmitter carried by the bird, can provide a fair estimate of metabolic rate over periods of a day or more (Owen 1969; Gessaman 1973, 1980; Ferns, MacAlpine-Leny & Goss-Custard 1980). Unfortunately, heart rate is related not only to metabolic rate, but also to other variables, such as stress. There are also technical problems with its use on unrestrained birds in the field. A further method of estimating the DEB is to compile time/activity budgets in the field, and to combine these with laboratory measurements of metabol-

Table 2.1. *Daily energy intake as a multiple of basal metabolic rate (BMR) in feeding waders*

Species	Situation	Daily intake ($\times$ BMR)	Source
Oystercatcher (*Haematopus ostralegus*)	Captivity	4.3	Hulscher (1974)
Oystercatcher (*Haematopus ostralegus*)	Field	6.8	Smith (1975), using data from Davidson (1968) and Hulscher (1974)
Oystercatcher (*Haematopus ostralegus*)	Field	5.8	Goss-Custard (1977*b*)
Ringed Plover (*Charadrius hiaticula*)	Field	2.5 to 4.4	Pienkowski (1980*a*)
Grey Plover (*Pluvialis squatarola*)	Field	3.0 to 3.9	Pienkowski (1980*a*)
Dunlin (*Calidris alpina*)	Field	3.2 to 4.2	Worrall (1981)
Bar-tailed Godwit (*Limosa lapponica*)	Captivity	3.0	Smith (1975)
Bar-tailed Godwit (*Limosa lapponica*)	Field	5 to 5.5	Smith (1975)
Redshank (*Tringa totanus*)	Field	4.2 to 5.0	Goss-Custard (1977*b*)

ism during different activities (e.g. Verner 1965). Measures of heart rate increments over quite short time periods (40 s) provide estimates of the energy expenditure of small components of activity, such as individual paces and pecks (P. N. Ferns, J. D. Goss-Custard & I. H. MacAlpine-Leny, unpublished).

Until such methods can be developed further, the energy requirements of most free-living birds can be estimated only indirectly, by measuring the food intake and, if possible, subtracting that portion which remains unassimilated. In order to convert estimates of energy intake to the DEB, it is necessary to have some knowledge of the extent of egestion and excretion. Although these can be measured quite easily by difference trials in the laboratory, surprisingly little has been done to quantify them. Evans *et al.* (1979) suggested that waders may assimilate about 80–90 % of the food they ingest.

Generally, estimates of the DEB of free-living birds seem to be between 2 and 4 times the BMR (Ebbinge, Canters & Drent 1975), to which must be added the food not assimilated if an estimate of the quantity of food taken is required. Some published estimates of daily intake for waders are shown in Table 2.1. We shall now discuss the ways in which these sorts of estimates were obtained.

Estimating food intake during the day

The food intake per day is likely to vary between individual birds but, in this chapter, we will limit ourselves to estimating mean rates for a population of unmarked birds.

In order to estimate food intake, the following measures are required:

(i)   the time spent feeding;
(ii) species of prey taken;
(iii) rate of intake;
(iv) size of prey taken;
(v)   energetic content of each prey type.

These five aspects are considered, in turn, below.

Time spent feeding

The simplest measure involves a regular count of how many birds in the study area are feeding and how many not. The integration of these proportions over the daylight period, or tidal cycle, gives an estimate of the mean time spent feeding in that period. (Of course, it is possible to measure the time spent feeding by individually recognizable birds, but this is much more difficult and time-consuming (both in the catching for marking and in the watching).) For study of some species, a little time may be needed to decide whether each individual is foraging or not; while it is immediately obvious if a tactilely searching Dunlin is feeding, a plover foraging by a stand-and-wait method may have to be watched for several seconds.

The method assumes that foraging and non-foraging birds are equally conspicuous. However, foraging birds may be more visible than non-feeders in poor light simply because they are moving. It is also necessary to know whether birds move out of sight or elsewhere, e.g. to a roost, when they stop feeding. If so, this must be allowed for. At Teesmouth (and many other sites) Redshanks feed mainly in creeks, where they may be hidden from view, but usually form small sub-roosts on the open mud, whereas some Grey Plovers do the reverse. Behaviour may vary according to weather conditions; Grey Plovers at Teesmouth feed more in creeks in high winds than in calm conditions (Dugan 1981a; see also Chapter 8). Vegetation may also conceal birds. These problems may be allowed for in deriving unbiased estimates of the time spent feeding.

When observations from more than one day need to be combined, the variation between days in the time of the tidal cycle in relation to daylight must be allowed for, as feeding is usually affected by state of tide and may also be concentrated in daylight.

Prey species identification

In contrast to the traditional view (e.g. Bryant 1979), the range of prey taken by a wader species varies greatly according to the types available in an area (e.g. Pienkowski 1981a, 1982). A survey of these is of value in deciding which methods of study of diet are most appropriate for a given wader species in a given area.

Determination of prey of birds was formerly achieved by killing specimens (usually by shooting) and examining their gut contents. In waders this method can have serious problems, as digestion of prey is generally very rapid, and traces of different prey species vanish at very different rates. For example, Goss-Custard (1969) found that Redshank digested *Corophium volutator* faster than *Hydrobia ulvae*. In this case, it was possible to deduce a correction factor for gizzard contents, based on a sample of birds which had some food in both the oesophagus (where no digestion had occurred) and the gizzard (Greenwood & Goss-Custard 1970).

In the case above, both prey species contained some hard parts, although their resistance to the grinding in the gizzard differed. In other situations, even this type of correction is inapplicable. At Lindisfarne, the main prey of several wader species are soft-bodied annelid worms, with no recognizable hard parts which survive digestion. Even Grey Plovers which had been seen taking mainly these prey items in the field, and were shot doing so (and the gut contents removed and preserved immediately), were found to contain only very few identifiable fragments of those prey (Pienkowski 1981a, 1982).

Clearly, the examination of gut contents is of little value when the proportion of entirely soft-bodied prey in the diet is high, but can be useful for birds of the same species in different places. For example, Schramm (1978) found that all important prey of Grey Plovers in the Swartkops estuary, South Africa, were crustaceans or molluscs, with hard parts. Here, analyses of gut contents were of value, as they can be when the proportion of soft-bodied prey in the diet is not too large, e.g. Wetmore (1925), Sperry (1940), Puttick (1978), Hicklin & Smith (1979).

Prey determination may also be based on analysis of regurgitated pellets or faecal droppings. These can be collected without any need to kill birds, but suffer the same problems of bias in relation to differences in digestion time of different prey. Indeed, the bias may be more marked than in analysis of gut contents, as rejecta may show what is not eaten, rather than what is. However, the approach can be valuable and its application to a study of the diet of the Dunlin, which has defied other methods, is described later. The use of carefully controlled doses of emetics on captured animals, to cause regurgitation of prey, might provide more representative samples than do rejecta, again without requiring killing of the animals. However, capturing birds reasonably quickly after feeding might be difficult.

Checking the validity and reliability of visual observations of feeding shorebirds and measuring the rate of prey intake

Analysis of gut contents or products cannot generally give information on the rate at which prey are taken. To do this usually requires observations of feeding birds. These may also be required for the basic description of diet, especially if other methods fail.

Estimates of prey intake rates from visual observations require confirmation and, if necessary, use of correction factors. For example, Goss-Custard (1973, 1977*a*) tested whether what he considered to be swallows of small prey by Redshank (*Tringa totanus*) were really so. He did this by observing captive birds (at similar distances to those in the field) taking artificial prey, in a cage in which the total numbers of prey present were counted, before and after the observation period. Reliability checks were carried out in the field, with two observers simultaneously (but independently) recording the feeding movements of the same birds.

In the study of plovers at Lindisfarne, neither of these methods were suitable, the first because plovers have a foraging behaviour which is space-demanding and not conducive to experiments in captivity, and the second because the field observation technique involved dictation into a continually running tape recorder, thus making independence between two observers difficult. Further, confirmation of the identification of prey type, and not only food intake rate, was required. Therefore, colour ciné film was taken of birds feeding in the natural situation. The processed film was subsequently viewed at normal speed and information recorded as in the field, by speaking into a tape recorder. The same film was later analysed frame by frame, to identify prey taken. Many months elapsed between making and viewing the film, and between viewing and frame-by-frame analyses, to avoid memory retention of previous assessments. Film analysis at normal speeds gives a conservative assessment of the identification ability of the observer. Comparison of prey so observed with those recorded from frame-by-frame analysis demonstrated that the identification of thin worms by the method used in the field was valid, and that the recording of items as either 'peck' or 'successful peck' was conservative, in that most of these were in fact captures of thin worms (Pienkowski 1982). At most, only 9 % of Grey

Plover pecks and 12 % of Ringed Plover pecks were unsuccessful, with more likely figures being 2 % for each species. N. C. Davidson (unpublished) also used ciné photography to assess pecking rates of Bar-tailed Godwits, as these were too fast to count directly with accuracy.

Although the features of the prey item taken successfully by a plover could be categorized only as, e.g. 'worm, thin, red, 1/2 bill-height long', identification to species level was often possible because invertebrate sampling revealed that only a limited range of potential prey was present. Thus field observations, in combination with sampling of benthic invertebrates, can be a powerful means of describing the diet of shorebirds.

Observations of prey capture at close quarters often reveals features which can be used to identify prey at greater distances. For example, *Arenicola marina* is distinctive because of its size, shape and fairly inert appearance while hanging from a bird's bill, whereas ragworms (principally *Nereis diversicolor*) have a 'ragged' appearance and tended to writhe rhythmically while suspended from the bill. Crabs have a distinctive appearance and birds deal with them in a characteristic manner: they hold the crab by each leg in turn and shake to detach this from the body before swallowing, and usually finish by swallowing the main body. Thin worms are taken by plovers by one or more delicate upward pulls, followed by a rapid flick of the bird's head as the worm comes free. Plovers feeding on thin worms often show fine dark lines across their breasts where the muddy worms have lashed. Brearey (1982) also used the method of capture to distinguish between worms and crustaceans taken by Sanderlings (*Calidris alba*).

Measuring the sizes of prey taken

In cases in which prey are identified from gut contents, pellets or faeces, it is nearly always possible to measure the size of a distinctive hard structure, and relate this to body size or weight by regressions calculated from intact animals collected from the field. A summary of structures suitable for measurement has been provided for numerous types of prey by Goss-Custard (1973).

Several possible biases must be borne in mind. For example, the antennal length of *Corophium* is a useful index of size, but its relation to total body size differs markedly according to sex and age (Boates & Smith 1979). Size itself may influence the time for which a particular individual is recognizable in the gut. The behaviour of the predator may differ when dealing with prey of different sizes. Evans (1975) found, in

Australia, that Pied Oystercatchers (*Haematopus ostralegus*) swallowed bivalves of less than 2 cm in length whole but extracted the flesh from larger individuals. Therefore, shells of the latter would not be found in the gut, and those of the former not among those discarded on the shore.

When estimating the size of prey taken during field observations of feeding birds, length of worms or width of hard-bodied prey have been assessed in relation to the bill length of the wader, or to the height of the bill from the ground. Such measures are amenable to the same sort of testing as those used for prey identification, such as ciné photography. Dugan (1981*a*) used a colleague on the mudflat, holding worms with forceps of the same length as that of the birds' bills, to check the accuracy of his estimation. When making such tests on elastic animals, it is important to bear in mind the way in which birds normally take the prey. A worm hanging loosely from a bill will appear different in length from the same worm when it is being pulled from its burrow.

Measuring the energetic content of prey

The number of different types of prey ingested is often a poor indication of their real contribution to the diet, because of their different sizes and water and energy contents (Table 2.2). Animals are best dried in a freeze-drier, or in a vacuum oven at 60 °C, since some volatile components may be lost at higher temperatures. In the case of bivalve molluscs, it is normally the calorific value of the flesh alone which is of interest, and so this needs to be removed from the shells. In the case of gastropods, removal of the shell may be quite difficult.

Apart from differences in dry matter and calorific content due to differing species and size, differences have been found in some situations relating to sex, age and season of the year (Goss-Custard *et al.* 1977; Boates & Smith 1979; papers reviewed by Dugan 1981*a*; see also below). These should be borne in mind when calculating calorific intake rates.

Estimation of food intake during the day by Dunlin in the Severn Estuary

A study was carried out between November 1978 and September 1980 at Penarth Flats, Cardiff, and at Collister Pill, near Magor (Worrall 1981). Here, Dunlin feed almost exclusively on three species – *Nereis diversicolor*, *Macoma balthica* and *Hydrobia ulvae*. Several species of oligochaete worms are extremely abundant at these sites, but do not contribute significantly to the diet, even though they do at Teesmouth (Evans *et al.* 1979).

Table 2.2 *Percentage contributions by numbers (and by energy content of various prey to the diet of plovers at Lindisfarne, Northumberland*

	Grey Plovers	Ringed Plovers
Whole *Arenicola*	1.2 (63)	0.1 (13)
Arenicola tails	0.4 (1.1)	0.1 (1.2)
Ragworm (*Nereis*)	0.04 (0.2)	0 (0)
Crabs (*Carcinus*)	0.1 (5.4)	0 (0)
Thin worms (*Notomastus, Scoloplos*)	78 (29)	65 (75)
Small prey	20 (1.3)	35 (11)

All three of the dominant prey species have hard parts which resist digestion, can be used to count the number of prey present, and can be measured to determine the size of prey from which they originated. *N. diversicolor* has a pair of chitinous jaws. For *M. balthica*, the minimum thickness of the valve alongside the hinge of the surviving fragments can be measured, using a microscope with an eyepiece micrometer. *H. ulvae* shells sometimes survive intact. Partially intact shell whorls are far more common, but it is impossible to deduce the absolute size from these. However, there is a good relationship between the biomass of flesh and the width of the operculum. Although these three measurements are all well correlated with the size and biomass of the respective organisms from which they came, there were significant differences in these relationships from month to month.

Three sources of material were examined for food remnants – faeces, the contents of the gizzard and oesophagus of birds killed under licence and pellets. The bulk of the material was voided orally as pellets. Faeces were at first collected singly, but when it was realized how little material they contained, bulk samples of 10 or 20 droppings were taken. These faeces were cleared with polyvinyl lactophenol, and then searched for fragments when spread out in water in a petri dish.

At first, pellets proved extremely difficult to find. In this area Dunlin produce large numbers of pellets during each day's feeding, but these are often poorly formed and cease to be produced shortly after feeding has stopped. It is impossible to collect many pellets on the main feeding areas (because pellets are widely dispersed and disintegrate too quickly), or on the main roost sites, because birds do not arrive there until too long after they have ceased feeding. The most effective sites for pellet collection are 'pre-roosts'. These form, progressively as the flats are covered, at or near the water's edge, and may merge with the high-tide roost during neap tides. Judging the time at which to disturb the pre-roost and search the area for pellets is quite critical, since the number of pellets increases with time, but the period available for searching the area before the tide covers it decreases. The number of Dunlin pellets collected between February 1978 and May 1980 was 920.

The pellets of all waders vary in size and appearance, depending upon the diet, so that identification of species from pellets alone is not possible. Droppings and, on certain substrates, footprints can be used to aid identification. In the Severn Estuary the average lengths and widths of pellets were 8.8×5.7 mm for Dunlin, 13.0×7.2 for Knot (*Calidris canutus*), 18.2×9.8 for Redshank and 25.4×18.3 for Curlew (*Numenius arquata*). In mid-winter most Dunlin pellets were golden yellow in colour, from ground-up *H. ulvae* shells. In spring and autumn, they were darker, and contained more silt and small stones.

From the faeces no whole *Hydrobia ulvae* shell whorls or opercula were recovered, even when a large proportion of the faecal material consisted of *H. ulvae* shell fragments. *Nereis diversicolor* jaws were present in droppings throughout the year, albeit in small numbers. The number of *Macoma balthica* hinges was small, even when the bulk of the faeces consisted of *M. balthica* shell fragments. Faeces did not provide a good guide to relative numerical abundance of prey, nor did they give a correct indication of the frequency distribution of size of prey taken, presumably because only the smaller fragments passed beyond the gizzard. The size distributions of prey fragments in gizzards and pellets showed no significant differences.

The dry weight of material in a pellet was considerable (monthly averages lying between 167 and 267 mg) and was highest in winter. The pellet material represented additional weight to be carried in the gut, and must also have impeded the transport and digestion of softer materials in the gut. The dry weight of material in the gizzard (monthly averages lying between 276 and 437 mg) was more variable than that in a pellet, presumably because birds were captured at differing times after feeding, some having already ejected some material.

There were marked monthly variations in the size distribution of all three prey species in the gizzards. These differences would not have been apparent if much indigestible material had remained in the gizzard for prolonged periods. Indeed, the turnover is so rapid that retention of prey fragments is probably not a serious bias in field studies where a high proportion of prey contains hard parts, even though birds in captivity may retain such fragments for long periods (Smith 1975).

Gizzards and pellets provided a good estimate of the size distribution of these prey, despite the tendency for measurable fragments of the smaller *M. balthica* and *N. diversicolor* to be lost in the faeces, simply because the number of fragments lost in droppings was so low. Dunlin produce droppings at a rate of approximately 28 per hour when feeding in mid-winter, so that 5 %, at most, of individuals of these two prey species that are ingested are lost via the droppings. Furthermore, the size distribution of these prey can be determined, provided that large enough samples of droppings are collected.

Unfortunately, these methods do not provide a good estimate of the relative proportions of the three prey species ingested, mainly because the opercula of *H. ulvae* are not as resistant as the hinges of *M. balthica* and the jaws of *N. diversicolor*. However, it is relatively simple to assess by eye the percentage volume of a pellet, gizzard or dropping occupied by shell fragments of each of the two molluscs. This is because *M. balthica* fragments are whitish, while *H. ulvae* fragments are dark brown. Over the size ranges taken by Dunlin, the shell weight to flesh ratio was almost identical within a month (but differed between months). Thus the percentage volume of shell fragments represented a good measure of the relative flesh intake of the two species.

There is no way of assessing the intake of *N. diversicolor* on a volumetric basis, since only their jaws remain intact. However, both the jaws of *N. diversicolor* and the hinges of *M. balthica* can be counted. The ratio of the two species found in pellets can be compared with the ratio observed being taken by Dunlin in the field. There was fairly good agreement between pellets and direct observations in November, December, January, February and April, but a poor one in October, March and May (Worrall 1981), when direct observations were more difficult.

Thus the relative numbers of the three species of prey ingested were estimated by counting, in pellets, jaws/2 for *N. diversicolor* and hinges for *M. balthica*, and determining the relative proportions of *M. balthica* and *H. ulvae* from the volume of shell fragments of each. Knowing the size ranges taken, and their relative shell and flesh weights, intake was then converted to numbers or biomass. The results are discussed in detail elsewhere. When expressed as biomass, *N. diversicolor* was the most important prey item, constituting at least half of the diet through-out the season. The rest of the biomass was made up of equal portions of *M. balthica* and *H. ulvae*. Combining the known feeding rates on *N. diversicolor* in February with the number of hours of daylight during which birds fed, and the relative composition of the diet deduced from pellets, suggests a food intake for Dunlin at this time of year of about

46 kcal per daylight period. This intake was achieved by the consumption of about 2100 individuals of *N. diversicolor*, 170 of *M. balthica* and 2100 of *H. ulvae* (Worrall 1981).

Seasonal patterns of energy intake during daylight hours

In winter in the temperate zone, temperatures are generally lower than at other seasons, winds stronger and precipitation greater. These conditions not only lead to a greater energy requirement for maintenance of body temperature, but also make food more difficult to obtain, leading to lower feeding rates. This is because, in cold conditions, intertidal invertebrates become less active; also, many bury deeper in the substrate and may move into the sub-littoral zone (e.g. Goss-Custard 1969; Smith 1975; Evans 1976, 1979; Pienkowski 1980*a*, 1981*a,b*; see also Chapters 1 and 4). Prey densities may also have been lowered significantly from autumn levels by predation (e.g. Evans *et al.* 1979).

If birds feed for only as long as they need to, one would expect an increase in the time they spend feeding each day in winter, and, indeed, this is generally observed (e.g. Fig. 2.1). Fig. 2.1 also shows that, in

Fig. 2.1. Percentages of time spent feeding, in daylight, by Ringed Plovers at Lindisfarne, Northumberland, in relation to season. Squares, neap tides; solid circles, spring tides; open circles, juveniles during spring tides. Where sufficient data are available, more than one point is given for each month and tide category. In such cases, the longer feeding time generally occurred on days of lower temperatures and higher winds (from Pienkowski 1982).

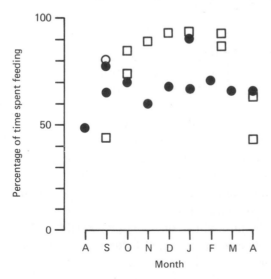

winter, Ringed Plovers generally fed for longer during neap tides than spring ones (because the main feeding areas were covered over the high water period of spring tides). In autumn and spring this difference did not occur, suggesting that feeding time was not limiting total daily intake then.

In Fig. 2.2, the estimates of feeding time and rates of energy intake of Ringed Plovers at Lindisfarne are combined to give estimates of intake per daylight period. In this example, the increase in feeding time in mid-winter is clearly inadequate even to compensate for the fall in intake rate, before allowing for the increased requirements. Intake in mid-winter, especially during spring tides, is much lower than in autumn or spring (when there appears to be little problem). It is also lower than the intake rate of more than 3 times the BMR expected by extrapolation from studies on energy requirements of a wide range of other birds (Ebbinge *et al.* 1975). Indeed, 4 times the BMR was found in the Dunlin study already described. In Grey Plovers at Lindisfarne the situation in

Fig. 2.2. Seasonal variations in daylight calorific intakes, expressed as number of BMR equivalents (from Pienkowski 1982). Squares, neap tides; solid circles, spring tides; open circles, juveniles during spring tides.

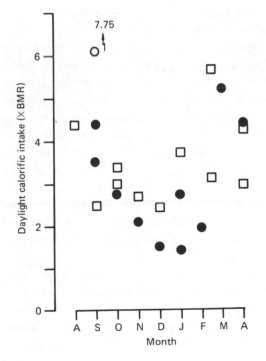

mid-winter appears to be even worse than in Ringed Plovers, estimates falling to < 1 BMR (Pienkowski 1982). These figures refer to daylight feeding only: what happens at night?

Nocturnal feeding

It has long been known that waders of many species feed at night in some circumstances, and assumed that feeding at night is less profitable than feeding by day (e.g. Evans 1976). Certainly, in some situations waders feed at night mainly when they cannot fulfil their requirements by day. Thus nocturnal feeding by plovers at Lindisfarne is largely confined to the winter months (Pienkowski 1982).

If nocturnal feeding is mainly used to 'top-up' daytime intake, what feeding rates are required at night?

We try to estimate this below, using the study of Ringed Plovers at Lindisfarne as an example (with numerical values given in parentheses), and taking the situation in January, when conditions should be near their least favourable for the birds.

Daily metabolic requirements (48.3 kcal) were estimated for a bird with a lean weight of 62 g for a given ambient temperature (3 °C) from relationships calculated empirically from a wide range of bird species of different weights (Kendeigh *et al.* 1977). The calorific intake in daylight at this temperature, estimated in the field at Lindisfarne (29.2 kcal in spring-tide conditions), has then been subtracted from these, to estimate the feeding requirements at night (19.1 kcal), if the energy budget is to be balanced in each 24 hours. These requirements have then been converted to the minimum calorific intakes required per minute at night (29.1 cal/min), assuming that feeding takes place for as long as intertidal feeding areas are exposed (maxima of 890 min on neap tides and 656 min on spring tides, respectively). Finally, these calorific intake rates have been converted to the number of worm-equivalents, assuming different sizes of worms (4.5 worms of 0.25 bill height/min or 2.0 worms of 0.75 bill height/min), and can be compared with pecking rates actually observed at night (7.3/min at 3 °C in February). For Grey Plovers, similar calculations give 13.8 worms of 0.25 bill height/min or 5.7 worms of 0.75 bill height/min, compared with an observed nocturnal pecking rate of 4.1/min at 3 °C in February.

The estimated required minimum feeding rates are thus of the same magnitude as the observed pecking rates. As there is clearly a number of errors which may attach to the estimated values, more detailed comparison would be unwise. It seems probable that in mid-winter well over half the energy intake of plovers may be obtained at night (see also Dugan 1981*b* for Grey Plovers).

It has often been assumed that the abilities of waders to detect cues given by prey, or to locate suitable sites for tactile foraging, are less at night than by daylight. This seems particularly likely in birds like plovers, which depend almost entirely on visual foraging. Their rate of pecking is depressed in some circumstances at night, and the distance moved to take prey (probably a measure of the range at which prey can be seen) generally so (Pienkowski 1983). However, many invertebrates are more active at the surface by night than by day (e.g. *Talitrus saltator*, Geppetti & Tongiorgi 1967; *Eurydice pulchra*, Jones & Naylor 1970; *Bathyporeia pelagica*, Fincham 1970a,b, Preece 1971; *Phyllodoce maculata*, Vader 1964, Pienkowski 1980a; *Nereis virens* and *N. diversicolor*, Dugan 1981b; *Carcinus maenas* and various other rocky shore animals, Naylor 1958, Kitching & Ebling 1967).

The antagonistic effects of decreased foraging ability but increased prey availability at night mean that the benefits of nocturnal foraging (if any) may vary considerably between different places and conditions. Seasonal variations may also occur if, for example, nocturnal activity of prey animals is related, at least in part, to periods of reproductive activity. Nocturnal foraging may even yield more biomass per minute, on average, than daylight foraging.

Dugan (1981b) noted that the importance of nocturnal foraging has been underplayed in some studies because, if estimates of daily requirements are found to lie close to the estimated intake in daytime, there is less incentive to investigate nocturnal foraging.

Dunlin on the Severn Estuary definitely fed at night, but a Twiggy Mark II image-intensifier did not produce sufficient magnification (3.5×) or discrimination for feeding rates to be measured (Worrall 1981). The rate of pellet production at dawn pre-roosts was not significantly less than at pre-roosts following prolonged daylight feeding, but their composition was different, with the number of *N. diversicolor* jaws depressed relative to the amount of *M. balthica* and *H. ulvae* shell fragments. Furthermore, nocturnal high-tide roosts (which could be observed quite efficiently with the image-intensifier) were always occupied for longer than the equivalent daytime roosts. These facts indicate that Dunlin probably obtained a smaller biomass in total by night than by day.

Clearly, nocturnal observations of foraging require a great deal of further work. The suitability of night-viewing apparatus (either infrared light sources or image-intensifiers) is rapidly increasing, but some types of equipment lack portability and many have too low a magnification or discrimination between plumage and sediment for ornithological studies.

Radio-tagging has been used to identify differences in diurnal and nocturnal foraging behaviours of individual Grey Plovers, as discussed later in this volume (Chapter 8). Luminous tags might also be useful although the effects of their presence would need investigation, especially if predation ·is important at night.

What period is relevant for energy balance?

Are energy intake and requirements balanced over each 24 hour period? We have implied that food intake during the low water feeding period must cover energy requirements also over the high water roosting period, and that deficits at night can be balanced by day and vice versa. This picture needs qualification.

The fortnightly cycle of spring and neap tides leads to variations in tidal height, in the duration of tidal emersion, and in the timing of tide in relation to daylight. One effect of those on energy intake of Ringed Plovers has already been illustrated (Figs. 2.1 and 2.2). A lunar monthly cycle occurs in nocturnal illumination, and often in the heights of tides, since one set of spring and neap tides in each cycle is usually more extreme than the other. Finally, an annual cycle in feeding conditions recurs, as outlined in an earlier section and in Chapter 1. In considering the period over which energy intake and expenditure are balanced, we must first examine the nature and use of energy reserves, which are usually stored on a periodic basis.

Functional significance of fat reserves

Waders, in common with most other birds, store energy primarily as fat, which yields over twice as much energy per unit weight (9.2 kcal/g) as other energy sources, such as protein or carbohydrate. A mid-winter peak in fat reserves of most wader species, in the parts of their range where harsh weather occurs, is well established (Davidson 1981b). However, acquisition of fat incurs costs, because of the energetic requirements of extra foraging, synthesizing the fat, carrying it while foraging, and probably also because of the risks associated with reduced flying ability (see Dick & Pienkowski 1979, Davidson 1981b). The amount of fat carried is an adaptive 'insurance policy' balancing the costs of carrying with the risk that such reserves are required (Evans & Smith 1975; Pienkowski, Lloyd & Minton 1979; Dugan *et al.* 1981; Davidson 1981a,b, 1982). Accordingly, fat reserves carried by waders in the north temperate zone decrease after mid-winter, as the risk of short periods of poor feeding conditions decreases. The quantity of fat carried appears to be adjusted, in Dunlin, to the average temperature of the particular wintering area (and presumably, therefore, to the chance of cold spells in that area) (Pienkowski *et al.* 1979, and see Fig. 2.5*b*). The size of the fat reserves appears to be adapted also to the foraging methods employed by a species: birds like plovers, which depend for successful detection of prey on the prey's activity, carry more fat than those tactile foragers that depend on prey being near the surface. This relates to depth distribution being generally less temperature-sensitive than activity (Pienkowski 1981*a*,*b*). Fig. 2.3 gives some examples of cycles of fat storage during the non-breeding season in northeast England, where waders regularly face harsh winter weather.

Protein also is stored as a nutrient reserve by waders, but is used as an energy source only when fat reserves are exhausted, during prolonged harsh weather (Evans & Smith 1975; Davidson 1981*a*) or long-distance, non-stop migrations (e.g. Dick & Pienkowski 1979).

What constitutes 'harsh' weather probably differs between species (and perhaps even between individuals). For example, during a week of freezing weather in northeast England in February 1978, fat reserves were not used by Bar-tailed Godwits (*Limosa lapponica*) or adult Dunlins, but were used by Golden Plovers (*Pluvialis apricaria*) and some first-year Dunlins (Davidson 1981*a*). This variation between species implies that fat reserves are used only as a last resort, when

Fig. 2.3. Fat reserves of waders wintering at Teesmouth and Lindisfarne, northeast England (from Evans & Smith 1975 and Davidson 1981*a*,*b*). Lipid index = mass of fat/total body mass. '——', adult Dunlins; '———', first-year Dunlins; '–·–', adult Grey Plovers; '····', adult Bar-tailed Godwits.

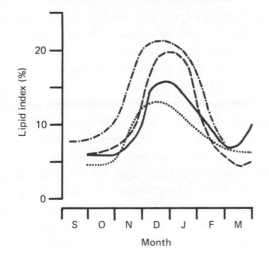

increased energy requirements cannot be balanced by a reduction in expenditure and/or an increase in intake. Because first-year Dunlins use more fat than adults during harsh weather, they should store more fat as an 'insurance' against such weather, and they do (Fig. 2.3). The study of changes in fat reserves can be used to identify the occasions when waders cannot balance their budget, except by drawing on internal reserves.

To detect such changes, the normal seasonal cycles of fat reserves, such as occur during mild winters, must also be known.

Assessing the size of fat reserves

Fat reserves are most conveniently expressed as a lipid index, usually the mass of fat expressed as a percentage of the total body mass (including fat). This measure is appropriate for the calculation of flight ranges, but does not allow direct comparison of survival times of waders of different body sizes. Because large birds have a lower metabolic rate per unit mass than small birds, the former can survive on stored fat for longer than can smaller birds with the same lipid index.

The amount of fat stored by waders has usually been determined accurately from samples of dead birds. Carcasses are oven-dried, preferably *in vacuo* at 60 °C, and fat extracted using a solvent, such as chloroform or petroleum ether (see Evans & Smith 1975; Davidson 1981b). As birds must be dead before their fat can be extracted in this way, changes in the amount of fat carried by individuals cannot be followed. However, it is usually assumed that a change in the mean levels of fat between samples can be interpreted as a change in the amount of fat stored by resident individuals. This can be confirmed by comparing the cycle of change in fat reserves determined from periodic samples with the weight changes of live individuals (an example is described later), and by ensuring that samples contain only individuals that are known to be resident, i.e. those ringed or marked on the same estuary earlier during the same non-breeding season.

Waders, like many other birds, lose weight rapidly after they have been caught (Lloyd, Pienkowski & Minton 1979; Davidson 1981b) through emptying of the gut and, subsequently, use of reserves, and as a result of stress induced by capture and handling. Weight lost after capture may not be regained for at least a week after release (Page & Middleton 1972; OAG Münster 1976). Hence weight changes in individual waders collected less than a week apart should be excluded from any calculations.

An alternative to carcass analysis for determination of fat levels is to estimate the fat reserves of live birds. In principle, this allows changes in the fat reserves of individuals to be examined. However, visual estimation and scoring of fat levels (in the abdomen, over the pectoral muscles and between the clavicles) is of limited use on individuals, because one score covers a range of lipid indices, so that only major changes in the amount of stored fat can be detected. Nevertheless, fat scores can be converted to lipid indices, using correlations of fat score and lipid index for each species derived from samples fat-extracted in the laboratory (e.g. McNeil 1969). The scale on which fat scores are assessed has not been standardized in the past. In waders, published studies have used scores of, e.g., 0–4 (McNeil 1969) and 1–5 (Prater 1975). A new standard code from 1 to 6 has been agreed for wader studies in many areas (Greenwood & Pienkowski 1978; Pienkowski 1980c).

An alternative to visual estimation of fat levels is measurement of the thickness of a fat layer, using a small ultrasonic probe. This technique has been used successfully on some waterfowl (Baldassarre, Whyte & Bolen 1980), but has yet to be tested on waders. One difficulty in using this technique on live birds may be to replicate accurately the position and angle at which the fat thickness is measured. As for a visual fat score, the fat thickness can be converted to a lipid index by using a killed sample on which values were measured.

Most changes in the total body weights of waders are caused by changes in the amount of stored fat (Johnston & McFarlane 1967; Dugan *et al.* 1981; Davidson 1981b), so the total body weight can also be used to estimate fat reserves. There are two main approaches.

(1) Calculation of an estimated lean (= fat-free) weight from correlations between lean weight and measures of body size, such as wing length and bill length (McNeil & Cadieux 1972; Page & Middleton 1972; Mascher & Marcstrom 1976; Davidson 1981b, 1983), followed by subtraction of the estimated lean weight from total body weight to give the weight of fat. The accuracy of estimation of lean weight is often sufficient only for the calculation of the mean lipid index of a sample, and not for that of individuals. Because lean weight varies seasonally and geographically, and because first-year waders are smaller and lighter than adults (Davidson 1981b), lean weight should be estimated only from a formula derived from a single age-class sample from the same season and area.

(2) Calculation of changes in total body weight of individual waders that have been weighed at least twice during the same (not different) winters. These individual weight changes are then compared with the seasonal cycle of fat reserves derived from periodic samples (e.g. Fig.

2.4). Individual changes in weight can be converted to changes in lipid index by assuming that each bird carried the average lipid index for that species/age/place on the date that it was first weighed. Assuming that all of a weight change is fat, the lipid index at any subsequent weighing can then be calculated as a proportion:

$$\text{lipid index } 2 = \frac{(\text{total body weight } 2 - \text{estimated lean weight})}{\text{total body weight } 2}$$

where

estimated lean weight = total body weight 1 − (total body weight 1 × lipid index 1)

(1 and 2 refer to first and second (or subsequent) weighings).

Changes in lipid index of individual Grey Plovers at Teesmouth, calculated in this way, correspond closely to the average seasonal cycle (Fig. 2.4b), showing that these individuals had not used their fat reserves as an energy source more than is usual at Teesmouth. This technique offers the best way of identifying the conditions under which individual waders cannot balance their energy budget without using fat reserves.

How long can fat reserves last? During severe weather, waders are usually found dying of starvation after about 7 days (Clark 1982; Davidson & Evans 1982), but starved Oystercatchers (*Haematopus*

Fig. 2.4. Seasonal changes in fat reserves of individual Grey Plovers at Teesmouth weighed twice during the same non-breeding season. Each solid line joins records of one bird, (*a*) of its total body weight, and (*b*) of its lipid index (see text). '——' shows lipid indices calculated from mean weights of netted samples (excluding individuals weighed twice) (from Fig. 2.3).

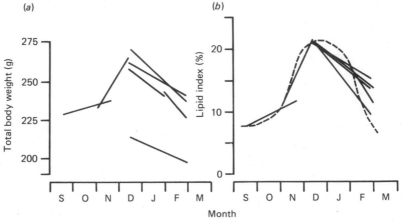

ostralegus) are sometimes found after only 3 days of severe weather (Heppleston 1971; Clark 1982). The exact time that fat reserves will last depends for each wader on the size of its reserves, its energy requirements and its food intake. Formulae relating metabolic rate and temperature (Kendeigh *et al.* 1977) predict that the mid-winter fat reserves of waders in northeast England are sufficient for survival, when the mean air temperature is 0°C, for 2–4 days by sandpipers and 4–8 days by plovers. This example assumes that there is no wind, and that the birds do not feed, so that all their energy requirements have to come from stored fat. During windy weather, heat loss is increased, so energy requirements are higher, and fat reserves will last for a shorter time. Conversely, even during severe weather, most waders can find some food, so only part of their energy requirement has to come from fat reserves. Consequently, energy balance may need to be considered over periods well in excess of a week.

Not all waders can regulate the size of their fat reserves throughout their wintering range. A wader that does not do so shows year to year variation in the timing and size of the mid-winter peak in fat, and in the size of the late winter reserves, depending on the severity of the weather in each year. Failure to regulate fat reserves results from continually inadequate food intake, particularly after mid-winter (Pienkowski *et al.* 1979; Davidson 1981*b*). Because there is no 'normal' size of fat reserves, the use of fat to balance energy requirements must be examined separately in each year.

Although most waders regulate their fat reserves over most or all of their wintering ranges, some Redshanks wintering in eastern Britain seem unable to do so (Davidson 1982). Just as Dunlins wintering in Britain store most fat where average weather conditions are most severe (Fig. 2.5*b*), so the mid-winter weights of Redshanks in southern and western Britain are highest on estuaries with the lowest average temperatures (Fig. 2.5*a*). However, the mid-winter weights on estuaries in eastern Britain are below the predicted regulated levels (Fig. 2.5*a*), indicating that less fat is stored than is needed as insurance against harsh weather. Indeed, the deficit in the fat reserves is greatest where average weather conditions are most severe. However, on at least some east coast estuaries, for example the Firth of Forth, weight changes of individual Redshanks during mild winters showed that part of the wintering population can regulate the amount of fat stored (Davidson 1982). After mid-winter, the fat reserves of Redshanks in eastern Britain decline more than in other waders: by late February, even in mild winters at Teesmouth, Redshanks had an average of only 2 % fat, compared with 5–8% in other waders (Davidson 1981*b*). This is probably sufficient for less than one day's survival from fat reserves alone, and Redshanks do often suffer higher mortality during severe weather than most other waders (Ash 1964; Dobinson & Richards 1964; Pilcher 1964; Pilcher, Beer & Cook 1974; Goss-Custard *et al.* 1977; Davidson 1981*a*; Clark 1982).

In addition to using fat reserves, Redshanks in eastern Britain also use muscle protein reserves during late winter (Davidson 1981*a*, 1982), so their lean weight decreases. This illustrates a more general point: that a decrease in total body weight can provide an overestimate of the amount of fat utilized, since part of the weight loss is of lean weight. It is not easy to find out what proportion of a weight change is lean weight. Preliminary methods of estimating the size of protein reserves and lean weight in live birds, such as that described by Davidson (1979), need further development. The seasonal cycle of total body weight follows a similar pattern (mid-winter peak followed by a decline), both when

Fig. 2.5. Total body weights of (*a*) Redshanks (from Davidson 1982) and (*b*) Dunlins (from Pienkowski *et al.* 1979) in relation to air temperatures on east coast (open circles) and south and west coast (solid circles) British estuaries in mid-winter. Each point shows a mean monthly weight (for Redshanks standardized for variation in body size – Davidson 1982) for one estuary ±1 S.E. Solid lines show in (*a*) the regression calculated from means from south and west coast estuaries only: total body weight of Redshank = 197.7 − 6.55 temperature, $r = -0.70$, $P < 0.05$; in (*b*) the regression calculated from means from all estuaries: total body weight of Dunlin = 73.6 − 3.33 temperature, $r = -0.80$, $P = 0.001$.

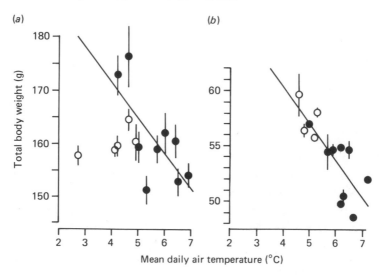

reserves are regulated and when non-regulated (Davidson 1981*b*). Absence of regulation is indicated by differences in the extent of the late winter decline from year to year and by the use of protein reserves even during mild winters.

In waders that regulate their fat reserves the peak amount of fat carried depends on how much is likely to be needed to balance energy requirements: those species or populations that face the greatest, most frequent, or prolonged deficits in energy balance store the most fat (Pienkowski *et al.* 1979; Davidson 1981*a,b*). Similarly, within a species that regulates its reserves, we would predict that most fat should be stored by those individuals that are least proficient in supplying their energy requirements from feeding, i.e. those with the lowest rates of energy intake, perhaps juveniles. However, some individuals of other species, e.g. Redshanks, probably can never feed fast enough to allow accumulation of adequate fat reserves. In such species, birds with the lowest rates of food intake will have small, but unregulated, reserves, whilst individuals with the highest rates of intake may also have small reserves, because they do not need larger ones. The relationship between energy intake and fat reserves has not yet been examined in individuals within a species, but only between age groups (Davidson 1981*b*).

Annual cycle adaptations

This chapter has been concerned with related problems of how birds balance their energy budgets and how research workers may be able to audit the accounting. We conclude by summarizing some of the ways in which birds may adjust components of their budget and, therefore, some of the points which may require attention in a study of them.

In difficult times, birds may minimize energy-demanding activities of two types: foraging and maintenance activities. It may be possible to adjust the foraging method, substrate, timing in relation to tide or daylight cycle, location, or prey selectivity, to reduce the energetic costs of foraging. Some effects of doing this were modelled by Evans (1976). In the longer term, activities requiring extra expenditure of energy, such as breeding, migration and moult, may have their timings adjusted in the annual cycle. For example, Pienkowski *et al.* (1976) argued that the duration and timing of wing moult is adjusted to coincide with periods of higher resource availability and to minimize the rate of utilization of resources within such periods.

If the timing of energy-demanding periods in the year cannot be adjusted (for example, the occurrence of harsh conditions in winter), energy reserves may be provided in advance. These normally take the form of subcutaneous fat deposits, and the size of such reserves may be adjusted as described earlier. Provision of external overwinter food reserves may also be possible, by behavioural methods. Some territorial Grey Plovers conserve stocks of prey by excluding conspecifics from an area from late autumn onwards (Dugan 1982, and Chapter 8 of this volume).

A further possibility which could allow waders to balance their energy budget more easily is movement between several sites during the non-breeding season, as conditions vary at these sites. The popular concept of movements in direct response to hard weather seems to apply less to coastal waders than inland species (Evans 1976; Davidson 1981*a*; Townshend 1982; Pienkowski & Evans 1983). However, patterns of movement adapted to the average conditions or to the probability of difficult conditions seem feasible, and large-scale movements within the winter certainly occur (e.g. Dugan 1981*a*; Pienkowski & Pienkowski 1981; Evans 1981). Investigations are in progress to describe these movements and consider the underlying reasons for them (Evans 1980; Pienkowski 1980*b*; Pienkowski & Pienkowski 1981; Pienkowski & Evans 1983).

Acknowledgements

The studies used as examples in this chapter have received financial support from the British Ornithologists' Union, the Natural Environment Research Council, and the Nature Conservancy Council. Unpublished data and valuable discussions have been provided by Dr P. J. Dugan, Dr P. R. Evans, Ann Pienkowski, F. L. Symonds and Dr D. J. Townshend. A chapter involving four authors and three editors is an unstable entity, and we are grateful to Ann Pienkowski for fighting the word-processor on our behalf.

References

Ash, J. S. (1964). Observations in Hampshire and Dorset during the 1963 cold spell. *Br. Birds* **57**: 221–41.

Baldassarre, G. A., Whyte, R. J. & Bolen, E. G. (1980). Use of ultrasonic sound to estimate body far deposits in the Mallard. *Prairie Nat.* **12**: 79–86.

Boates, J. S. & Smith, P. C. (1979). Length–weight relationships, energy content and the effects of predation on *Corophium volutator* (Pallas) (Crustacea: Amphipoda). *Proc. Nova Scotia Inst. Sci.* **29**: 489–99.

Brearey, D. M. (1982). Feeding ecology and foraging behaviour of Sanderling *Calidris alba* and Turnstone *Arenaria interpres* at Teesmouth, north-east England. Unpublished PhD thesis, University of Durham, South Road, Durham DH1 3LE, England.

Bryant, D. M. (1979). Effects of prey density and site character on estuary usage by overwintering waders (Charadrii). *Est. coast. mar. Sci.* **9**: 369–84.

Bryant, D. M. & Westerterp, K. R. (1980). The energy budget of the House Martin (*Delichon urbica*). *Ardea* **68**: 91–102.

Byrkjedal, I. (1975). Skeletal remains of microrodents as a source of calcium for Golden Plovers *Pluvialis apricaria* during the egg-laying period. *Sterna* **14**: 197–8. (In Norwegian with English summary.)

Clark, N. A. (1982). The effects of the severe weather in December 1981 and January 1982 on waders in Britain. *Wader Study Group Bull.* **34**: 5–7.

Davidson, N. C. (1979). A technique for protein reserve estimation in live Redshank *Tringa totanus*. *Wader Study Group Bull.* **27**: 14–15.

Davidson, N. C. (1981*a*). Survival of shorebirds (Charadrii) during severe weather: the role of nutritional reserves. In *Feeding and survival strategies of estuarine organisms*, ed. N. V. Jones & W. J. Wolff, pp. 231–49. Plenum Press, New York.

Davidson, N. C. (1981*b*). Seasonal changes in the nutritional condition of shorebirds (Charadrii) during the non-breeding seasons. Unpublished PhD thesis, University of Durham, South Road, Durham DH1 3LE, England.

Davidson, N. C. (1982). Changes in the body condition of Redshanks during mild winters: an inability to regulate reserves? *Ringing & Migration* **4**: 51–62.

Davidson, N. C. (1983). Formulae for estimating the lean weight and fat reserves of live shorebirds. *Ringing & Migration* **4**: 159–66.

Davidson, N. C. & Evans, P. R. (1982). Mortality of Redshanks and Oystercatchers from starvation during severe weather. *Bird Study* **29**: 183–8.

Davidson, P. E. (1968). The Oystercatcher – a pest of shell-fisheries. In *The problems of birds as pests*, ed. R. K. Murton & E. N. Wright, pp. 141–55. Academic Press, London.

Dick, W. J. A. & Pienkowski, M. W. (1979). Autumn and early winter weights of waders in north-west Africa. *Ornis Scand.* **10**: 117–23.

Dobinson, H. M. & Richards, A. J. (1964). The effects of the severe winter of 1962/63 on birds in Britain. *Br. Birds* **57**: 373–474.

Dugan, P. J. (1981*a*). Seasonal movements of shorebirds in relation to spacing behaviour and prey availability. Unpublished PhD thesis, University of Durham, South Road, Durham DH1 3LE, England.

Dugan, P. J. (1981*b*). The importance of nocturnal foraging in shorebirds: a consequence of increased invertebrate prey activity. In *Feeding and survival strategies of estuarine organisms*, ed. N. V. Jones & W. J. Wolff, pp. 251–60. Plenum Press, New York.

Dugan, P. J. (1982). Seasonal changes in patch use by a territorial grey plover: weather-dependent adjustments in foraging behaviour. *J. anim. Ecol.* **51**: 849–57.

Dugan, P. J., Evans, P. R., Goodyer, L. R. & Davidson, N. C. (1981). Winter fat reserves in shorebirds: disturbance of regulated levels by severe weather conditions. *Ibis* **123**: 359–63.

Ebbinge, B., Canters K. & Drent, R. (1975). Foraging routines and estimated daily food intake in Barnacle Geese wintering in northern Netherlands. *Wildfowl* **26**: 5–19.

Evans, P. R. (1975). Notes on the feeding of shorebirds on Heron Island. *Sunbird* **6**: 25–30.

Evans, P. R. (1976). Energy balance and optimal foraging strategies: some implications for their distributions and movements in the non-breeding season. *Ardea* **64**: 117–39.

Evans, P. R. (1979). Adaptations shown by foraging shorebirds to cyclical variations in the activity and availability of their invertebrate prey. In *Cyclic phenomena in marine plants and animals*, ed. E. Naylor & R. G. Hartnoll, pp. 357–66. Pergamon Press, Oxford.

Evans, P. R. (1980). Movements of wader populations between estuaries: some questions raised by studies at Teesmouth. *Wader Study Group Bull.* **29**: 6.

Evans, P. R. (1981). Migration and dispersal of shorebirds as a survival strategy. In *Feeding and survival strategies of estuarine organisms*, ed. N. V. Jones & W. J. Wolff, pp. 275–90. Plenum Press, New York.

Evans, P. R., Herdson, D. M., Knights, P. J. & Pienkowski, M. W. (1979). Short-term effects of reclamation of part of Seal Sands, Teesmouth, on wintering waders and Shelduck. *Oecologia* **41:** 183–206.

Evans, P. R. & Smith, P. C. (1975). Studies of shorebirds at Lindisfarne, Northumberland. 2. Fat and pectoral muscles as indicators of body condition in the Bar-tailed Godwit. *Wildfowl* **26:** 37–46.

Ferns, P. N., MacAlpine-Leny, I. H. & Goss-Custard, J. D. (1980). Telemetry of heart rate as a measure of estimating energy expenditure in the Redshank *Tringa totanus* (L.). In *A handbook on biotelemetry and radio tracking*, ed. C. J. Amlaner & D. W. Macdonald, pp. 595–601. Pergamon Press, Oxford.

Fincham, A. A. (1970*a*). Amphipods in the surf plankton. *J. mar. biol. Ass. UK* **50:** 177–98.

Fincham, A. A. (1970*b*). Rythmic behaviour of the intertidal amphipod *Bathyporeia pelagica*. *J. mar. biol. Ass. UK* **50:** 1057–68.

Geppetti, L. & Tongiorgi, P. (1967). Nocturnal migrations of *Talitrus saltator* (Montagu) (Crustacea Amphipoda). *Monitore Zool. Ital.* (N. S.) **1:** 37–40.

Gessaman, J. A. (1973). *Ecological energetics of homeotherms*, vol 20. Utah State Press, Logan. 155 pp.

Gessaman, J. A. (1980). Heart-rate as an indirect measure of daily energy metabolism of the American Kestrel. *Comp. Biochem. Physiol.*, *A* **65:** 273–89.

Goss-Custard, J. D. (1969). The winter feeding ecology of the Redshank *Tringa totanus*. *Ibis* **111:** 338–56.

Goss-Custard, J. D. (1973). Current problems in studying the feeding ecology of estuarine birds. *Coast. Ecol. Res. Pap.* **4:** 1–33.

Goss-Custard, J. D. (1977*a*). Optimal foraging and the size selection of worms by Redshank, *Tringa totanus*, in the field. *Anim. Behav.* **25:** 10–29.

Goss-Custard, J. D. (1977*b*). The ecology of the Wash. III. Density-related behaviour and the possible effects of a loss of feeding grounds on wading birds (Charadrii). *J. appl. Ecol.* **14:** 721–39.

Goss-Custard, J. D., Jenyon, R. A., Jones, R. E., Newbury, P. E. & Williams, R. le B. (1977). The ecology of the Wash. II Seasonal variations in the feeding conditions of wading birds (Charadrii). *J. appl. Ecol.* **14.** 701–19.

Greenwood, J. J. D. & Goss-Custard, J. D. (1970). The relative digestibility of the prey of Redshank *Tringa totanus*. *Ibis* **112:** 543–4.

Greenwood, J. J. D. & Pienkowski, M. W. (1978). Transferring wader ringing data between computers. *Wader Study Group Bull.* **22:** 17–19.

Hails, C. J. & Bryant, D. M. (1979). Reproductive energetics of a free-living bird. *J. anim. Ecol.* **48:** 471–82.

Heppleston, P. B. (1971). The feeding ecology of Oystercatchers (*Haematopus ostralegus* L.) in winter in northern Scotland. *J. anim. Ecol.* **40:** 651–72.

Hicklin, P. W. & Smith, P. C. (1979). The diets of five species of migrant shorebirds in the Bay of Fundy. *Proc. Nova Scotia Inst. Sci.* **29:** 483–8.

Hulscher, J. B. (1974). An experimental study of the food intake of the Oystercatcher *Haematopus ostralegus* L. in captivity during the summer. *Ardea* **62:** 155–70.

Johnston, D. W. & McFarlane, R. W. (1967). Migration and bioenergetics of flight in the Pacific Golden Plover. *Condor* **69:** 156–68.

Jones, D. A. & Naylor, E. (1970). The swimming rhythm of the sandbeach isopod *Eurydice pulchra*. *J. exp. mar. Biol. Ecol.* **4:** 188–99.

Kendeigh, S. C., Dol'nik, V. R. & Gavrilov, V. M. (1977). Avian energetics. In *Granivorous birds in ecosystems*, ed. J. Pinowski & S. C. Kendeigh, pp. 127–204. Cambridge University Press, Cambridge, England.

Kitching, J. A. & Ebling, F. A. (1967). Ecological studies at Lough Ine. *Adv. Ecol. Res.* **4:** 197–291.

Lasiewski, R. C. & Dawson, W. R. (1967). A re-examination of the relation between standard metabolic rate and body weight in birds. *Condor* **69:** 13–23.

Lloyd, C. S., Pienkowski, M. W. & Minton, C. D. T. (1979). Weight loss of Dunlins while being kept after capture. *Wader Study Group Bull.* **26:** 14.

Maclean, S. F. (1974). Lemming bones as a source of calcium for arctic sandpipers (*Calidris* spp.). *Ibis* **116:** 552–7.

McNeil, R. (1969). La détermination du contenu lipidique et de la capacité de vol chez quelques espèces d'oiseaux de rivage (Charadriidae et Scolopacidae). *Can. J. Zool.* **47:** 525–36.

McNeil, R. & Cadieux, F. (1972). Numerical formulae to estimate flight range of some North American shorebirds from fresh weight and wing length. *Bird Banding* **43:** 107–13.

Mascher, J. W. & Marcstrom, V. (1976). Measures, weights and lipid levels in migrating Dunlins *Calidris a. alpina* L. at the Ottenby Bird Observatory, South Sweden. *Ornis Scand.* **7:** 49–59.

Naylor, E. (1958). Tidal and diurnal rhythms of locomotor activity in *Carcinus maenas* (L.). *J. exp. Biol.* **35:** 602–10.

OAG Münster (1976). Zur Biometrie des Alpenstrandläufers (*Calidris alpina*) in den Rieselfeldern Münster. *Vogelwarte* **28:** 278–93.

Owen, R. B. (1969). Heart rate, a measure of metabolism in blue-winged teal. *Comp. Biochem. Physiol.* **31:** 431–6.

Page, G. & Middleton, A. L. A. (1972). Fat deposition during autumn migration in the Semi-palmated Sandpiper. *Bird Banding* **43:** 85–96.

Pienkowski, M. W. (1980a). Aspects of the ecology and behaviour of Ringed and Grey Plovers *Charadrius hiaticula* and *Pluvialis squatarola*. Unpublished PhD thesis, University of Durham, South Road, Durham DH1 3LE, England.

Pienkowski, M. W. (1980b). WSG co-operative project on movements of wader populations in western Europe. *Wader Study Group Bull.* **29:** 7.

Pienkowski, M. W. (1980c). A new WSG data form and prospects for analysis. *Wader Study Group Bull.* **28:** 11–14.

Pienkowski, M. W. (1981a). Differences in habitat requirements and distribution patterns of plovers and sandpipers as investigated by studies of feeding behaviour. *Verh. orn. Ges. Bayern* **23:** 105–24.

Pienkowski, M. W. (1981b). How foraging plovers cope with environmental effects on invertebrate behaviour and availability. In *Feeding and survival strategies of estuarine organisms*, ed. N. V. Jones & W. J. Wolff, pp. 179–92. Plenum Press, New York.

Pienkowski, M. W. (1982). Diet and energy intake of Grey and Ringed Plovers, *Pluvialis squatarola* and *Charadrius hiaticula*, in the non-breeding season. *J. Zool., Lond.* **197:** 511–49.

Pienkowski, M. W. (1983). Changes in the foraging pattern of plovers in relation to environmental factors. *Anim. Behav.* **31:** 244–64.

Pienkowski, M. W. & Evans, P. R. (1984). Migratory behavior. In *Shorebirds*, ed. J. Burger & B. L. Olla, *Behaviour of marine animals*, vol. 6. Plenum Press, New York.

Pienkowski, M. W., Knight, P. J., Stanyard, D. J. & Argyle, F. B. (1976). The primary moult of waders on the Atlantic coast of Morocco. *Ibis* **118:** 347–65.

Pienkowski, M. W., Lloyd, C. S. & Minton, C. D. T. (1979). Seasonal and migrational weight changes in Dunlins. *Bird Study* **26:** 134–48.

Pienkowski, M. W. & Pienkowski, A. (1981). WSG project on movements of wader populations in western Europe: second progress report. *Wader Study Group Bull.* **31:** 16–17.

Pilcher, R. E. M. (1964). Effects of the cold weather of 1962/63 on birds of the north coast of the Wash. *Wildfowl Trust Ann. Rep.* **15:** 23–6.

Pilcher, R. E. M., Beer, J. V. & Cook, W. A. (1974). Ten years of intensive late-winter surveys for waterfowl corpses on the north-east shore of the Wash, England. *Wildfowl* **25:** 149–54.

Prater, A. J. (1975). Fat and weight changes of waders in winter. *Ringing & Migration* **1:** 43–7.

Preece, G. S. (1971). The swimming rhythm of *Bathyporeia pilosa* (Crustacea: Amphipoda). *J. mar. biol. Ass. UK* **51**: 777–91.

Puttick, G. M. (1978). The diet of the Curlew Sandpiper at Langebaan Lagoon, South Africa. *Ostrich* **49**: 158–67.

Schramm, M. (1978). The feeding ecology of Grey Plover on the Swartkops estuary. Unpublished BSc (Hons.) thesis, Dept of Zoology, University of Port Elizabeth, South Africa.

Smith, P. C. (1975). A study of the winter feeding ecology and behaviour of the Bar-tailed Godwit *Limosa lapponica*. Unpublished PhD thesis, University of Durham, South Road, Durham DH1 3LE, England.

Sperry, C. C. (1940). Food habits of a group of shore birds: Woodcock, Snipe, Knot, and Dowitcher. *Wildlife Res. Bull. Washington* **1**: 1–37.

Townshend, D. J. (1982). The Lazarus syndrome in Grey Plovers. *Wader Study Group Bull.* **34**: 11–12.

Utter, J. M. & Lefebvre, E. A. (1973). Daily energy expenditure of Purple Martins (*Progne subis*) during the breeding season: estimates using D_2O^{18} and time budget methods. *Ecology* **54**: 597–603.

Vader, W. J. M. (1964). A preliminary investigation into the reactions of the infauna of the tidal flats to tidal fluctuations in water level. *Neth. J. Sea Res.* **2**: 189–222.

Verner, J. (1965). Time budget of the male long-billed marsh wren during the breeding season. *Condor* **67**: 125–39.

Wetmore, A. (1925). Food of American phalaropes, avocets, and stilts. *Bull. US Dept Agric.* **1359**: 1–20.

Worrall, D. H. (1981). The feeding behaviour of Dunlin *Calidris alpina* (L.). Unpublished PhD thesis, University College, Cardiff, UK.

3

Relations between the distribution of waders and the intertidal benthic fauna of the Oosterschelde, Netherlands

P. MEIRE & E. KUYKEN

Introduction

All over the world, intertidal land in coastal areas and estuaries is disappearing rapidly, as a result of reclamation schemes and other threats. Such changes have already occurred in the 'Delta' area in the southwest of the Netherlands (Fig. 3.1), where closure of the Haringvliet and the Grevelingen estuaries in the early 1970s caused important losses of intertidal marshes and mudflats. At present, the Oosterschelde estuary is also subject to considerable ecological changes, by enclosure of two areas and by reduction in tidal amplitude of the remaining semi-open part of the estuary (see Chapter 15 of this volume). The 'Delta' area in total is of outstanding importance for wildfowl and wader populations during a great part of the year. The conflict between coastal engineering and birds in the region has been discussed by Saeys & Baptist (1977, 1980).

Until recently, little was known about the quantitative effects of habitat loss on the bird populations of estuaries. Conservation planning requires information both on habitat use and on factors affecting the numbers of birds on these wetlands. This type of information, for two British estuaries, is to be found in Goss-Custard (1977a, 1979), Evans *et al.* (1979) and Evans (1981).

The study area and methods used

Detailed ecological research on the bird numbers and their distribution in the changing area of the Oosterschelde started in 1979. Our study area, called 'Slikken van Vianen', is situated on the south coast of Schouwen-Duiveland, in a side branch of the Oosterschelde (Fig. 3.1).

It consists of a small saltmarsh (30 ha) and 510 ha of intertidal flats. Here, the sediment is mainly sand, except for some lower parts with more silt where mussel beds (*Mytilus edulis*) occur. Some zones are covered with *Zostera marina*, *Zostera noltii* and *Enteromorpha* sp. The salinity of the water ranges between 28–30 g/litre.

Beginning in 1976–77, monthly, and from 1981 fortnightly, counts of waders and waterfowl were made at high-water roosts on the saltmarsh or adjacent agricultural land.

In 1979, 14 permanent plots (1.0 or 0.5 ha each) were staked out on the mudflat at different levels and on different substrate types, each in a homogeneous zone. The number and the activity of birds was noted in these 14 plots every 30 min during most of a whole tidal cycle, but during the first and the last hour in which each plot was exposed, counts were made every 15 min. This was done on 18 days between August and December 1979.

In September 1979, five substrate samples were taken in each plot (156 cm² area, 35 cm in depth). After fixation, they were sieved through a 1-mm² mesh. Macrobenthic organisms were identified, counted and weighed. Individuals of important prey species were also measured, in

Fig. 3.1. Map of the Delta area and location of the study site 'Slikken van Vianen'.

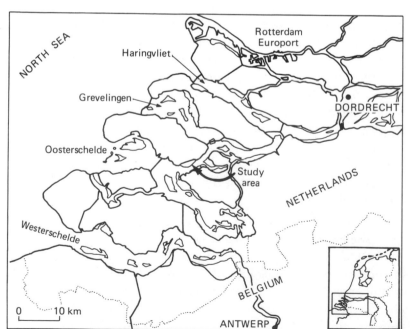

order to determine length/weight regressions. Finally, soil composition of the plots was also analysed, for grain distribution (cf. Wolff 1973), and amount of organic matter.

Wader numbers at the Slikken van Vianen
Waders use different parts of the Oosterschelde in different ways. Thus it is necessary to describe fluctuations in numbers of the most important species specifically in our study area (Fig. 3.2).

Three species, Grey plover (*Pluvialis squatarola*), Curlew (*Numenius arquata*) and Dunlin (*Calidris alpina*) use the area chiefly during autumn

Fig. 3.2. Monthly average numbers (for the period 1976–81) of six wader species on the Slikken van Vianen.

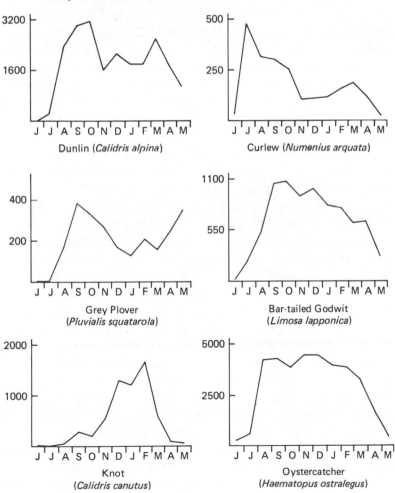

and spring migration. Knot (*Calidris canutus*) are most abundant in winter, as in the eastern part of the Oosterschelde (Chapter 15 of this volume). Oystercatcher (*Haematopus ostralegus*) numbers remain steady from August till January and then decline gradually. Bar-tailed Godwits (*Limosa lapponica*) are most numerous in September, declining slowly thereafter.

Five-year average patterns can hide important variations from year to year, caused especially by weather conditions. However, during the observation period in 1979, bird numbers generally corresponded with these averages, except that Oystercatchers reached maximum numbers in December.

Relations between the presence of birds and the benthos

It is well known that zonation of benthic invertebrate species exists on mudflats (see e.g. Anderson 1972; Wolff 1973). Some years ago, Wolff (1969) showed that in the Delta area six regions could be distinguished on the basis of their infauna. He also distinguished six groups of waders, each with a characteristic distribution, covering one or more invertebrate regions. The wader distributions depended chiefly on the distribution of their principal prey species but also on other environmental factors.

In our analysis of data from the 14 plots in a very small part of the Delta area, we were also able to divide the waders into different groups associated with their usage of the intertidal flats.

We examined first the degree of similarity between the plots (using the Canberra metric and group average sorting techniques). The results (Fig. 3.3) are presented as dendrograms (Clifford & Stephenson 1975), based upon biomass (or numbers) of the benthic organisms and upon bird feeding minutes (or wader densities). Notwithstanding the high overall similarities between the plots, we can distinguish four clusters, which are nearly the same whether based upon birds or benthos. These clusters are listed in Table 3.1, alongside the average total biomass density of Mollusca and Annelida in each. The proportion of biomass contributed by each species of invertebrate within each phylum is shown in Fig. 3.4. Table 3.2 summarizes the average occurrence of wader species in the clusters, expressed as feeding minutes/0.1 ha and as feeding densities/ha. This table is based upon counts of complete tidal cycles during 5 days between 26 and 31 August 1979.

The different clusters of plots (Fig. 3.3) are situated in different zones of the intertidal flats. At the highest levels, those of cluster IV have a rather low biomass density, made up by only a few invertebrate species.

Bird density is also very low and the area is used mainly on the ebbing tide, for example by Bar-tailed Godwits. Many waders pass quickly through this zone while feeding along the falling waterline. When the flood tide reaches the zone, the birds no longer need to feed; hence their overall presence in these plots is low and chiefly involves four species (Table 3.2).

The lower in the intertidal area the plots are situated, the higher the biomass density of invertebrates they contain. The plots in cluster III contain few important species of benthos, but a higher biomass than those of cluster IV. Six species of waders use the area on both ebbing

Fig. 3.3. Similarities between 14 sample plots presented as dendrograms. (*a*) based upon benthic biomass (g dry weight/0.1 m²); (*b*) based upon use by waders (feeding minutes/0.1 ha).

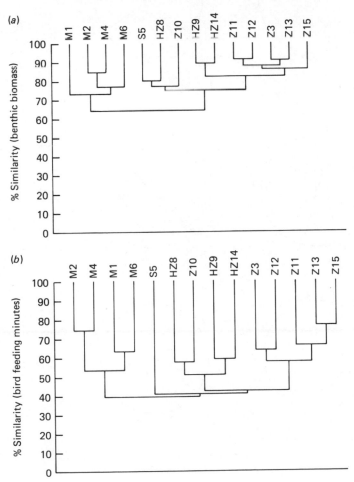

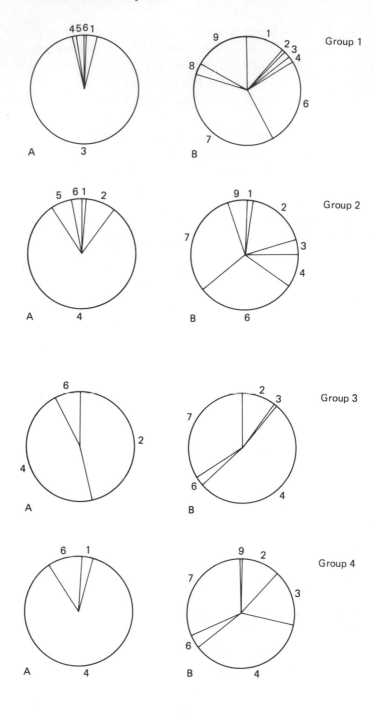

Fig. 3.4. Average proportional composition of Mollusca (A) and Annelida (B) in each cluster of plots. (A) 1, *Littorina littorea*; 2, *Hydrobia ulvae*; 3, *Mytilus edulis*; 4, *Cerastoderma edule*; 5, *Scrobicularia plana*; 6, *Macoma balthica*. (B) 1, Oligochaeta: 2, *Nereis* sp.; 3, *Nephtys hombergii*; 4, *Scoloplos armiger*; 5, *Capitella capitata*; 6, *Heteromastus filiformis*; 7, *Arenicola marina*; 8, *Ampharete acutifrons*; 9, *Lanice conchilega*.

and flooding tides, at lower densities but feeding for a greater number of bird-minutes than in the plots of cluster IV (Table 3.2).

The plots at the lowest tidal levels are characterized by very rich benthic communities. Cluster I (the mussel bed) is clearly distinct from the others. Cluster II (plots holding predominantly *Cerastoderma*) would be heterogeneous if classified by sediment or by bird usage. In particular, plot S5 is aberrant because of its exceptional use by Bar-tailed Godwits. Plots of both clusters I and II contain high benthic biomass densities and diversities, and birds are also very numerous. The density of many wader species remains almost constant there during the whole period that the surface is exposed.

There is a significant positive correlation between the numbers of bird species and benthic invertebrate species in the series of 14 plots ($r = 0.67$, $P < 0.01$). This suggests that the most 'productive' lower tidal levels of study area are also the most important for feeding waders. These results, however, are based upon observations at the end of August. From further research, there is good evidence that the higher parts of the mudflats become more important towards and during winter.

Feeding density of Oystercatchers on a mussel bed

Besides benthic biomass density, it became clear that other factors affected bird density. This was investigated for Oystercatchers feeding on *Mytilus edulis* on the four plots of cluster I (on plots of cluster II they took mainly *Cerastoderma edule*).

We found an exponential increase in the density (or feeding minutes/ha) of Oystercatchers with increasing *M. edulis* density (Fig. 3.5) in August and September; i.e. Oystercatchers showed a preference for the richest plots. In December, however, their density was almost even over the whole mussel bed, including the plots with lower biomass densities. (In this month, only two points are shown in Fig. 3.5 because the other two plots disappeared, but low-water counts of the mussel bed as a whole revealed similar overall densities of birds.)

The increase in numbers of Oystercatchers in the area, from 3000 in September to 6000 in December, and its effect on feeding densities, is

Table 3.1. *Average biomass of invertebrates and abiotic factors associated with four clusters of sample plots in the intertidal areas of the Slikken van Vianen*

Clusters:	I	II	III	IV
Plots (see Fig. 3.3)	M1/2/4/6	S5/HZ8/Z10	HZ9/14	Z11/12/3/13/15
Mollusca (total dry biomass, g/m^2)	3822	571	139	59
Annelida (total dry biomass, g/m^2)	49	23	16	04
Substrate	Muddy sand	Muddy sand	Muddy sand	Fine sand
Degree of sorting of particles	Well	Well	Well	Well–less well
Silt content (%)	11.9	11.0^a	2.3	3.8
Organic matter (%)	0.6	0.43	0.3	0.2
Exposure time	4 hours	5–7 hours	8 hours	4–5 hours
Plant coverage	—	$Zostera^b$	*Enteromorpha* sp.	—

[a] Except for S5, which was 5%.
[b] HZ8 only.
Crustacea (chiefly *Carcinus maenas*) were present only at biomass densities of less than $1 g/m^2$.

Table 3.2. *Occurrence of wader species in clusters of plots on the mudflat*

	Cluster I		Plot S5		Cluster II[a]		Cluster III		Cluster IV	
	FM	FD	FM	FD	FM	FD	FM	FD	FM	FD
Haematopus ostralegus	376	12.75	88	4.03	312	10.26	54	0.65	71	3.05
Numenius arquata	182	5.55	63	2.09	47	1.32	73	1.03	54	2.15
Limosa lapponica	62	2.47	667	31.61	36	0.93	59	0.60	61	2.77
Tringa totanus	51	2.03	0	00.00	240	1.58	7	0.06	0	00.00
Tringa erythropus	49	2.01	76	3.28	0	00.00	0	00.00	0	00.00
Tringa nebularia	3	0.15	0	00.00	2	00.00	0	00.00	0	00.00
Calidris alpina	123	4.94	208	8.90	60	2.10	89	1.38	72	2.59
Pluvialis squatarola	4	0.19	53	2.08	14	0.43	66	1.75	3	0.14
Arenaria interpres	54	1.20	0	00.00	47	1.30	54	1.05	0	00.00
Charadrius alexandrinus	0	00.00	0	00.00	5	00.10	16	0.25	4	0.15
Total	904	31.29	1155	51.99	763	18.02	418	6.77	265	10.85

FM = feeding minutes/0.1 ha.
FD = feeding densities/ha.
[a] Excluding plot S5.

Fig. 3.5. Relation between *Mytilus edulis* biomass density and
Oystercatcher feeding density in August, September, December 1979.
(Each point represents the data from a separate plot in cluster I.)

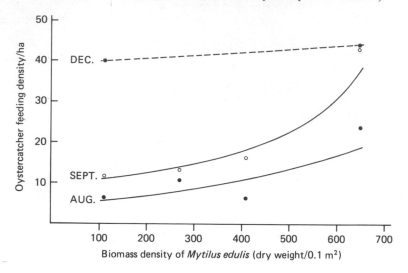

Fig. 3.6. Feeding density of Oystercatchers/ha in two plots, compared
with the increasing number in the study area (July–December 1979).

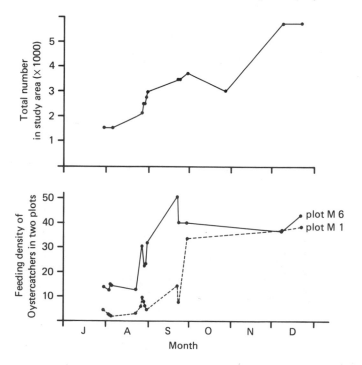

illustrated in Fig. 3.6. By mid-September the most productive parts of the mussel bed (e.g. plot M6) had been occupied, so that bird density there had reached a plateau. Further arrivals of birds on the study area led to increasing use of rather poorer or marginal parts of the mussel bed (e.g. plot M1), on which the same overall feeding density was reached later.

Zwarts (1974) and Goss-Custard (1977*a,b*), studying much larger areas, also found that with increasing numbers of birds the areas less favoured by the first arrivals became progressively more important. The results of the present work suggest that this principle also holds within a single mussel bed. Furthermore, the density of *feeding* Oystercatchers on the best plot (M6) did not increase as fast as the *total* density of Oystercatchers on that plot (Fig. 3.7). This implies again that birds should settle in the more marginal parts of the mussel bed when densities in the best food area become very high. Thus, not only benthic biomass density, but also behavioural factors affect the densities of Oystercatchers that are able to feed on a mussel bed. Behavioural mechanisms underlying this have been reviewed recently by Goss-

Fig. 3.7. Feeding density in relation to total (feeding + non-feeding) density of Oystercatchers on plot M6.

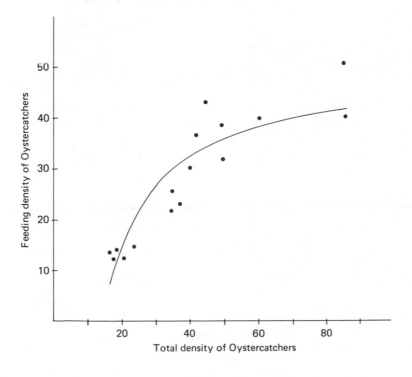

Custard (1980). Koene (1978) gives evidence for interference between birds: with growing Oystercatcher density, their feeding activity as well as food capture and intake rates decline. Vines (1980) also showed that the number of aggressive encounters in this species increased with decreasing distance between individuals. Perhaps behavioural limitations on bird density are most important in species whose invertebrate food is not mobile.

Acknowledgements

We thank H. Baptist and P. Meininger (Rijkswaterstaat Middelburg) for substantial help in the field and for placing wader counts at our disposal. We are also grateful to our colleagues A. De Kimpe and L. Vanhercke for discussing statistical matters. The first author acknowledges a grant as Aspirant of the Belgian Foundation for Scientific Research (NFWO).

References

Anderson, S. S. (1972). The ecology of Morecambe Bay II. Intertidal invertebrates and factors affecting their distribution. *J. appl. Ecol.* **9**: 161–78.
Clifford, H. T. & Stephenson, W. (1975). *An introduction to numerical classification.* Academic Press, San Francisco. 119 pp.
Evans, P. R. (1981). Reclamation of intertidal land: some effects on Shelduck and wader populations in the Tees estuary. *Verh. orn. Ges. Bayern* **23**: 145–61.
Evans, P. R., Herdson, D. M., Knights, P. J. & Pienkowski, M. W. (1979). Short-time effects of reclamation of part of Seal Sands, Teesmouth, on wintering waders and Shelduck. I. Shorebirds' diets, invertebrate densities and the impact of predation on the invertebrates. *Oecologia* **41**: 183–206.
Goss-Custard, J. D. (1977a). The ecology of the Wash. III Density-related behaviour and the possible effect of a loss of feeding grounds on wading birds (Charadrii). *J. appl. Ecol.* **14**: 721–39.
Goss-Custard, J. D. (1977b). Predator responses and prey mortality in Redshank *Tringa totanus* L., and a preferred prey, *Corophium volutator* (Pallas). *J. anim. Ecol.* **46**: 21–35.
Goss-Custard, J. D. (1979). Effect of habitat loss on the numbers of overwintering shorebirds. In *Shorebirds in marine environments*, ed. F. A. Pitelka pp. 167–77. Cooper Ornithological Society.
Goss-Custard, J. D. (1980). Competition for food and interference among waders. *Ardea* **68**: 31–52.
Koene, P. (1978). De Scholekster: aantalseffecten op de voedselopname. Dissertation, Zoology Department, University of Groningen, Netherlands.
Saeys, H. L. F. & Baptist, H. J. M. (1977). Wetlands criteria and birds in a changing delta. *Biol. Cons.* **11**: 251–66.
Saeys, H. L. F. & Baptist, H. J. M. (1980). Coastal engineering and European wintering wetland birds. *Biol. Cons.* **17**: 63–83.
Vines, G. H. (1980). Spatial consequences of aggressive behaviour in flocks of Oystercatchers, *Haematopus ostralegus* L., *Anim. Behav.* **28**: 1175–83.
Wolff, W. J. (1969). Distribution of non-breeding waders in an estuarine area in relation to the distribution of their food organisms. *Ardea* **57**: 1–28.
Wolff, W. J. (1973). The estuary as a habitat. *Zool. Verh. Leiden* **126**: 1–242.
Zwarts, L. (1974). *Vogels van het brakke getijgebied.* Bondsuitgeverij, Amsterdam. 212 pp.

4

How Oystercatchers and Curlews successively deplete clams

L. ZWARTS & J. WANINK

Introduction

The leading problem tackled in most studies of the feeding ecology of coastal wading birds is to what degree the food supply on the intertidal flats is a limiting factor for the bird populations (see review by Goss-Custard 1980). Their food supply – mainly macrobenthic animals living in the substrate – is highly variable from season to season and also from year to year. The study by Beukema (Beukema, Bruin & Jansen 1978; Beukema 1979) of the intertidal flats of the Balgzand (western Wadden Sea) shows that the spatfall of important shorebird prey species like the clam *Mya arenaria*, the mussel *Mytilus edulis* and the cockle *Cerastoderma edule* is very erratic. Their survival over the winter period also varies from year to year.

The variability in prey density is even larger if we take into account the fluctuations in the number of prey which are likely to be taken by the different wading birds; some of the prey are ignored because they are too small to be profitable, whereas others lie too deep in the substrate and are out of the reach of the waders' bills (Reading & McGrorty 1978).

We summarize in this chapter some results of current research on colour-banded birds feeding on the mudflats along the Frisian coast, near the village of Moddergat. We will explain why Oystercatchers (*Haematopus ostralegus*) and Curlews (*Numenius arquata*) feeding on clams select different size and depth classes and will show that both bird species combine to deplete the food stocks in the course of 2–3 years.

Selection of the profitable clams

Not all prey accessible to birds are in fact taken. For example, Herring gulls (*Larus argentatus*) ignore shore crabs (*Carcinus maenas*) below *c*. 20 mm carapace width, whilst the rejection threshold for the Curlew is *c*.10 mm, and for the Redshank (*Tringa totanus*) and Greenshank (*Tringa nebularia*) it is still lower (*c*. 5 mm) (Zwarts 1981). It was suggested by MacArthur & Pianka (1966) that the lower acceptance threshold was chosen by the bird in such a way that a maximal intake rate was ensured. Prediction of the lower limit for Oystercatchers feeding on mussels was correct (Zwarts & Drent 1981), and the data of Hulscher (1982) for Oystercatchers preying upon *Macoma balthica* also support the rule: the observed lower limit taken, of *c*.11 mm shell length, is near the expected acceptance threshold.

The applicability of this rule to our study of waders feeding on clams can be investigated by comparing the handling efficiency (mg ingested per second of handling) of each size class of prey with the overall feeding rate (mg ingested per second of feeding), where feeding = searching + handling. Optimal foraging theory predicts that prey sizes for which the handling efficiency is below the average rate of food intake should be ignored (see Krebs 1978 for a general review of the topic).

We observed that Oystercatchers feeding on clams did not take shells below a certain size. Since it is difficult to measure the prey size exactly in a field situation, we decided to capture some Oystercatchers and offer them clams of different size classes in controlled experiments, as Hulscher had done with *Macoma balthica*.

In one of the experiments an Oystercatcher fed on clams of different size classes buried in the substrate, all at the same depth, just below the surface. The yield per second of handling time appeared to be maximal for shells of 40–45 mm (Fig. 4.1).

Hulscher and his students made a series of observations in October and November 1980 on free-living Oystercatchers feeding on clams in our study area. They measured the handling of about 3000 prey and collected the fresh shells recently opened by Oystercatchers. They found that clams of 15–40 mm (average 28.2 mm) were taken (Fig. 4.2; left panel). But in contrast to the captive bird, the free-living Oyster-catchers did not take all the flesh from the shells. In many cases they pulled out the siphon, together with the greater part of the body, in one jerk. On average 22 % of the flesh remained in the shell and the flesh taken averaged 51 mg ash-free dry weight. The mean handling time was 5.83 s, so the average yield during handling was 8.74 mg/s (open square in Fig. 4.1c). This field estimate is quite close to the performance curve of the captive Oystercatcher.

Fig. 4.1. (*a*) Weight (g ash-free dry weight) of clam in relation to its size. (*b*) Handling time for clams of different size classes by a captive Oystercatcher (left) and a free-living Curlew (right). (*c*) Yield (mg ash-free dry weight) per second of handling time of clams of different size classes. For the Oystercatcher (left) this was determined in July 1981, whereas the handling efficiency of the Curlew (right) is derived from measurements in different months (August–November; 1978 and 1979 combined). The lines are drawn by eye. The lower acceptance threshold for clam size (solid triangle) is determined by the point where the extrapolated line meets the level of mean feeding rate (in the left panel the feeding rate of free-living Oystercatchers was used; J. Hulscher, unpublished). The open square and the open circle are discussed in the text. *Note:* the data from the Oystercatcher concern an experimental situation where all clams were present just below the surface, whilst the Curlew data were collected on the mudflats, where the big clams are living at greater depth (see Fig. 4.3). If depth itself has an effect on the handling efficiency, which seems likely, the slope for both species cannot be compared directly.

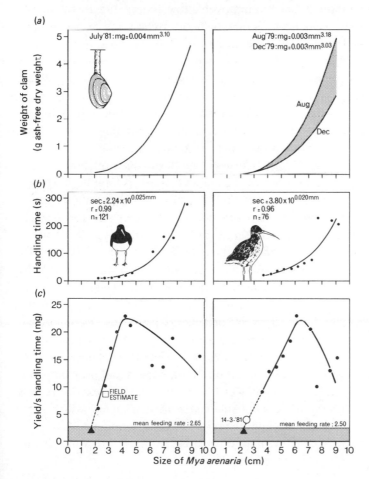

Since free-living Oystercatchers devote 30.3 % of their feeding time to handling the clams (the other 69.7 % being spent in searching for prey), the average rate of intake of biomass amounted to $0.303 \times 8.74 = 2.65$ mg/s (shaded block in Fig. 4.1c). Theory predicts that all prey sizes yielding a lower rate of biomass intake would be ignored. The smallest clam given to the captive bird was 22 mm. The handling efficiency for this is still above the predicted lower threshold. For clams between 20 and 40 mm there is a linear relationship between yield and shell size. Extrapolating the line downwards, we would expect that clams of 17.3 mm and smaller should be rejected, which fits well with the observed threshold in the free-living Oystercatchers (Fig. 4.2).

The clam is also an important prey for Curlew, but the threshold size required for acceptance is higher. For this species we measured the profitability of different shell sizes for free-living Curlews. It was easy to measure the handling time but impossible to estimate directly the size of the shells taken, because a Curlew pulls the siphon and pieces of flesh out of the shell, from below the mud surface. It was possible, however, to locate the clams after they had been eaten, for a Curlew makes a small crater while handling the prey. We were able to locate the clams seen to be taken in relation to a grid of thousands of numbered poles, separated from each other by a few metres on the feeding area. After an observation period, during which we timed the handling of the clams being eaten and noted their positions, we searched the different craters. To calculate the yield for the Curlews, the flesh which remained in the shell was weighed and subtracted from an estimated total flesh weight, derived from a weight/length relationship. Fig. 4.1 shows the handling time and the profitability as a function of the shell size for a single colour-banded Curlew (called 20Y), which fortunately feeds near one of our observation towers and has done so for five years. As Fig. 4.1 shows, large clams appear to be more profitable than small ones. Most clams taken by the bird were indeed large (Fig. 4.2a). The risk to a clam of being taken was maximal if it measured 4–5 cm (Fig. 4.2b), but the risk to the most common clam size present (2–3 cm) was 130 times less.

From Fig. 4.1c it can be seen that clams below 23.6 mm length would be unprofitable for this bird. In March 1981, when large clams were very rare, we observed that the marked bird took some small prey but since we could not locate the clams which were taken, their exact size was unknown. Assuming these were solely second-year clams, which averaged 25 mm in size at that time, the bird would have obtained 4.14 mg/s handling time (indicated with an open circle in Fig. 4.1c), which is just above the rejection threshold.

Certain other colour-banded Curlews also took clams of about 25 mm, or even smaller, but in those cases the birds swallowed the clam whole, including the shell. The handling time was very short: 3.41 s ($n = 252$) for two Curlews observed in November 1980; hence they obtained quite a high yield of 9.2 mg/s handling time. Eating clams in this way may have disadvantages, however, for it was observed predominantly in

Fig. 4.2. (*a*) Size selection by Oystercatcher (left) and Curlew (right), preying on clams (% frequency distribution per 5 mm class; sample size indicated). Clams eaten by free-living Oystercatchers in October–November 1980 or Curlew in August–December 1978 are compared to the clam population on offer in the Moddergat area for Oystercatcher and in the Nes area for Curlew. The right panel shows the prey selection by a single colour-banded Curlew (20Y); the size has been derived from the known relationship between handling time and weight (and thus size) of the clams. (*b*) Risk to a clam of being taken, relative to the maximal predation risk of a single size class. The maximal risk is set to 1 (Oystercatcher: clam size 3–4 cm; Curlew: clam size 4–5 cm). The relative risk is derived from the data given in the upper panel. The solid triangle denotes the predicted lower limit and the open triangle the most profitable size class (see Fig. 4.1).

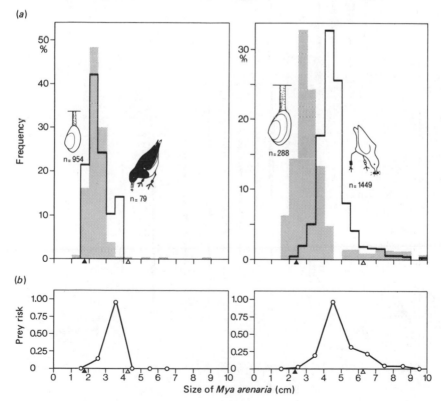

periods when big clams were very rare (the winter of 1977–78 and of 1980–81), and even in those lean winters most of the colour-banded Curlews for which clams were the main prey never swallowed the small ones. Perhaps they would have caused a 'digestive bottleneck' (Kenward & Sibly 1978).

Fig. 4.3. Frequency distribution of depth of clam per size class (0–4, 5–9 mm, etc.). The mean and the number of clams dug out are indicated. With the aid of grey bars the part of the clams being taken by Oystercatcher and Curlew are shown. The inset figure shows the accessible percentage per size class for Oystercatcher (upper 7 cm of the substrate), and for male and female Curlew (upper 12 and 14 cm, respectively).

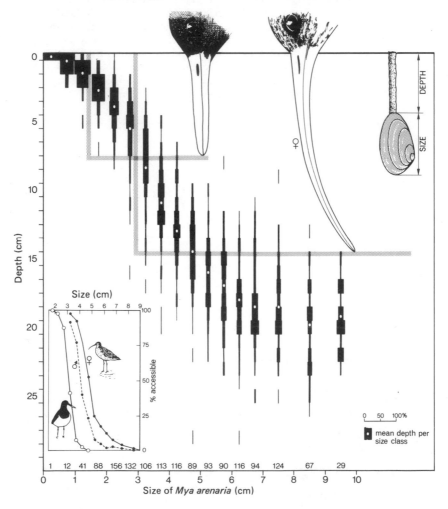

The accessible fraction of the profitable clams

The data on the yields of food from different sizes explain why Oystercatchers take smaller sizes than Curlews, but cannot help us to understand why the Oystercatchers on the mudflats do not attack successfully those clams of over 4 cm and why Curlew 20Y took relatively few clams above 6 cm (Fig. 4.2). These upper limits are in fact determined by the proportion of the different size classes within the reach of the bills of the two waders.

To investigate the depth distribution of the clams in each size class, we used a circular corer (of surface area $176 \, cm^2$) pushed 40 cm into the substrate. After breaking open the core, it is possible to measure accurately the distance between the mud surface and the upper tip of the bivalves. From these measurements it appears (Fig. 4.3) that an Oystercatcher cannot find clams of more than 4 cm in length in the upper 7 cm of the substrate (the bill length of an Oystercatcher) and that the greater part of the shells above 7 cm length are out of the reach of the female Curlew (bill length 14 cm). Even fewer are accessible to a male (bill length 12 cm) (Fig. 4.3, inset).

Most clams taken by the marked Curlew lived, given their size, at remarkably shallow depths (Fig. 4.4). Nearly all were buried above the mean depth of their respective size classes. The bird managed to find the very rare clams of more than 6 cm length which lived at atypically shallow depths, only 8–14 cm below the mud surface.

In the summer of 1979 we manipulated the food supply in the former feeding territory of the marked Curlew, by burying some hundreds of big clams just below the surface of the intertidal flats. As an unexpected but interesting result, an Oystercatcher started to prey on the big clams and vigorously defended the site, with success, not only against congeners but also against Curlews. It did so for a couple of days, during which period this single bird completely depleted the planted population of prey. This kind of behaviour was never observed in 'natural' situations. Interactions between Oystercatchers and Curlews are normally quite rare.

In the natural situation, it is very unlikely that there is direct competition between Oystercatcher and Curlew when they both feed on clams, because of the small overlap between the prey sizes which are available as well as profitable for both bird species (Fig. 4.3). Since Curlews select larger clams than Oystercatchers, there is a segregation in time if the two species are to feed on clams from the same year of spatfall.

However, this potential partitioning of resources does not tell us anything quantitatively about competition for food between the species. Oystercatchers might deplete the clam stock completely before the shells have reached the size at which they can be harvested profitably by Curlews.

Depletion of the food stock

Since we started our sampling programme of the macrobenthic fauna in 1977, there has been one successful spatfall of clam (1979); the previous one was in 1976.

Spatfall occurs during the summer. First-winter clams reach a size of *c.* 8 mm and are thus still too small to be utilized by Oystercatchers. During the second growing season most animals pass the lower acceptance threshold for Oystercatchers, but not until the next year do they become profitable for Curlews (Fig. 4.5*a*). During the growth of the shell, clams also bury deeper. The size–depth relationship (Fig. 4.3), can be used to derive the average depth of each cohort (Fig. 4.5*b*), but also allows calculation for all sampling dates of the number of clams above the lower acceptance size threshold and which are accessible to

Fig. 4.4. Depth of clams on offer in relation to their size, and depth of 84 clams taken by the colour-banded Curlew 20Y (dots). The grey field shows the depth range at which 95 % of the clams are living (based on the same data as in Fig. 4.3).

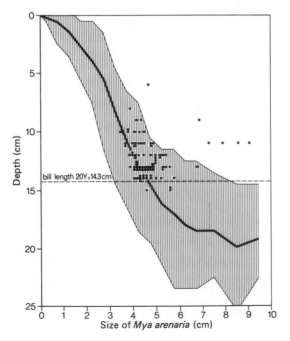

bill length 20Y=14.3cm

Fig. 4.5. (*a*) Average size of two cohorts of clams of which the spatfall took place in 1976 (open circles) and 1979 (solid circles) in the Nes area (Frisian coast); rejection threshold for both bird species is indicated. (*b*) Average depth of two cohorts of which the spatfall took place in 1976 (open circles) and 1979 (solid circles); bill length (and thus maximal prey depth) for both bird species is given. (*c*) Density of clams/m^2 since 1977 in the Nes area (number of two cohorts together). Upper panel shows total density and density of clams vulnerable to predation by Oystercatchers (clam size ≥15 mm, depth ≤7 cm, see Fig. 4.3). Lower panel gives density of potential prey for Curlews (size ≥30 mm, depth ≤14 cm, see Fig. 4.3). Note the different scale used for both bird species. Main exploitation periods (as derived from bird counts and observations on prey selection) are indicated by black bars along the x-axis.

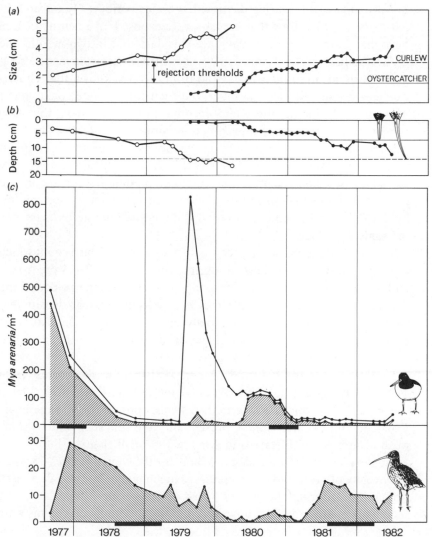

both bird species (Fig. 4.5c). Good years for clam-eating Oystercatchers (the winters of 1977–78 and 1980–81) precede the rich years for Curlews (the winters of 1978–79 and 1981–82).

The predation pressure by Oystercatchers and Curlews appears to be high enough to explain the greater part of the loss of clams after the second growing season.

In October and November 1980 all Oystercatchers present on the mudflats preyed upon clams. From the work of Hulscher we know that the average intake was 3.63 clams/min ($n = 870$ min). Because the water level was measured continuously, we know that the mudflats were exposed for 15900 min in daylight during these two months and for 21500 min at night. During daytime low-water periods, Oystercatchers fed for 88 % of the time that the flats were exposed, at a density of 1.26 birds/ha. Combining these data, we calculate that in two months the Oystercatchers took 6.4 clams/m^2 by day. Bird counts were made also at night, with infra-red binoculars, and from this it is known that Oyster-catchers remained to feed at night, but their feeding rate could not be measured. Assuming the same feeding rate as by day, the total predation in October and November amounted to 15.1 clams/m^2.

If the birds which were present during the rest of the winter continued to feed on clams and did not switch to the only alternative prey available, *Macoma balthica*, the resultant predation pressure in the period 1 October 1980 to 1 March 1981 can be estimated at 20 clams/m^2 by day only, or 49 clams/m^2 in total if the predation rate by night was the same as by day.

This value is a minimum, since we have omitted one bird count which took place just after a cold spell, during which the mudflats were frozen and many Oystercatchers fed on dying bivalves. During this period the density rose to 33 birds/ha, 20 times as high as the average density that winter. This situation lasted for between 3 and 5 days. In that short period as many as 20 clams/m^2 might have been eaten if the birds had achieved the same feeding rate as in autumn.

Between 1 October 1980 and 1 March 1981 there was a decrease from 110 to 20 clams/m^2 (Fig. 4.5c). At maximum, Oystercatchers took 69 of the 90 clams/m^2 which disappeared. A smaller proportion were eaten by Common Gulls (*Larus canus*), present after the cold spell, and by some of the Curlews which started to swallow the small clams.

No detailed observations are available for the winter of 1977–78 when the density of preferred size classes was much higher but also decreased dramatically during the winter (Fig. 4.5c). Counts of birds on the feeding area are available, however, and we also know that clams were the main (and perhaps the only) prey taken because we found many clams recently opened by Oystercatchers. The average Oystercatcher density in the winter of 1977–78 was 5.1 times as high as in the winter of 1980–81. Assuming the same intake rate by the birds, the predation pressure in the period October–March would have been 351 clams/m², assuming equally heavy predation by day and by night. The decrease which was found – from 440 to 120 clams/m² – was in fact below the estimated impact of the Oystercatchers.

The predation pressure by Curlew is more difficult to measure, since there is a large variation in the food choice of different individuals. The majority of male Curlews, for instance, never take clams, nor do some of the females. The best data at hand concern the predation by the marked female 20Y within her territory. The density of the accessible clams of the preferred size classes amounted to 8 clams/m² in July 1978. In the next nine months, she took at maximum about half of the food stock in her territory of 6000 m².

Discussion: prey risk and depth

The mortality rate of the clam decreases with age (Brousseau 1978). It is well documented that plaice exert heavy predation on the spat of the clam (Smidt 1951; de Vlas 1979). In our study area, bird predation on the spat is quite low. Knot (*Calidris canutus*) take them, but this wading bird was usually rather rare in our area.

The data presented above show that the lower acceptance threshold for the birds is determined by the yield in relation to handling time, whilst the upper threshold is set by the decreasing proportion of the clams of the larger classes living within the reach of the bill. Amongst the accessible clams there is a relatively heavy predation on the larger size classes, which are the most profitable prey. Oystercatchers take many clams of 3–4 cm in length, although most clams of that size class lie at depths beyond the Oystercatchers' bill. Clearly an essential part of the life strategy of the clam should be to grow a long siphon as quickly as possible so that it can live at a safe depth as soon as possible. The fact that the siphon weight of a small clam amounts to 50 % of its total body weight, compared to 30 % in large clams, attests to this. The selection by 20Y of clams which live at shallower depths than average for their respective size classes (Fig. 4.4), indicates that there is a lower risk to prey living at greater depths. The same has been found by Myers, Williams & Pitelka (1980) in experiments with crustacean prey of Sanderlings (*Calidris alba*).

Oystercatchers start to feed on clams during the bivalve's second

growing season as they pass the lower size limit of about 17 mm. A year
after that, most clams live beyond the reach of the bill of the Oystercatcher. How long individual clams are vulnerable to predation by
Oystercatchers depends on the growing/burrowing rate, which is highly
variable. A period of some weeks or months only is indicated (Fig.
4.6a). However, the differences between individual clams are so great
that the cohort remains exploitable for a year (Fig. 4.5). Curlews have a
longer period in which to take clams which still live within the reach of

Fig. 4.6. (*a*) Depth of six individual clams on an intertidal flat between
April and November 1981, measured with the aid of a thin nylon
thread attached on the shell. Size was determined in April and
November. Maximal probing depth for Oystercatchers is given. Note
the relationship between size increment in the three second-year clams
and the vulnerable period. (*b*) Changes in depth of three clams planted
at a shallow depth, to show that big clams have not lost the ability to
burrow.

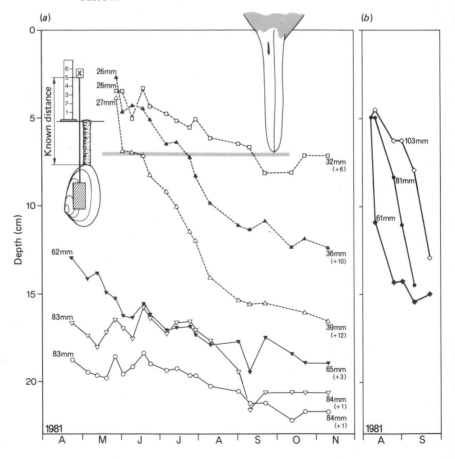

their bills, but by selecting only clams above 3 cm in length, they suffer a considerable reduction in the density of available prey. The number of clams which are buried just out of reach of the Curlew bill after they pass the acceptable size threshold is rather great. The biomass of preferred sizes of clams able to be reached by male Curlews with bills of 11–12 cm is only about half of that accessible to females (bill sizes 14–16 cm). This is probably one of the reasons why most males do not feed on clams, whilst most females do.

Clams which live below the reach of the Curlew bill (a maximum of 16 cm) are out of danger. Measurements of the depth of individual clams (Fig. 4.6) show that there is no short-term variation in their living depth. Yet it is possible that very big clams become accessible to avian predators if the upper layer of the substrate is eroded. Continuous measurements made on the variation in absolute height of the intertidal flats show that after heavy storms erosion of about 10 cm may occur. On the other hand, big clams are able to bury themselves again. In contrast to what is generally believed, big clams planted just below the mud surface are able to burrow *c.* 5 cm within a few days (Fig. 4.6*b*).

We are convinced that precise measurements on the accessible fraction of that part of the food supply which can be harvested profitably are necessary to understand the predator–prey relationships. Nature is complex, however. On most intertidal flats and also in our study area, wading birds seldom meet a one-prey situation. Normally, in attempts to maximize their rate of biomass intake, birds can decide what prey to take and thus where to go. In autumn 1980, for instance, most Oystercatchers did not remain on the mudflats, but switched to nearby cockle and mussel beds, which came into existence after the heavy spatfall of the year before. Switching was profitable because the intake rate on the mudflats, where the clam was the only prey, was lower than on the cockle beds (J. B. Hulscher, unpublished) and on the mussel banks (Zwarts & Drent 1981). Switching by short-billed male Curlews from mudflat-feeding to meadow-feeding in winter is another example (Townshend 1981). The substantial differences between the prey choice of individual colour-banded Curlews in our study area (B. Ens & L. Zwarts, unpublished) show that feeding decisions have to be studied even on the level of individual birds. Switching on the part of the individual predator to more profitable prey is thus another factor determining the predation risk to alternative prey. This makes us again aware of how much work still has to be done before it is possible to ascertain the role of food as a limiting factor for the large number of wading birds feeding on the intertidal flats.

Conclusions

Oystercatchers and Curlews ignore prey which are unprofitable, i.e. those of which the handling efficiency is below the intake rate during feeding (handling + searching) as predicted by optimal foraging theory (Figs. 4.1 and 4.2*a*).

The predation risk is maximal for clams which are about 1 cm above the lower acceptance threshold (Fig. 4.2*b*). Bigger clams are taken less often since they live out of reach of the bill (Fig. 4.3). Given a size class, clams which are buried less deep have a greater risk of being taken (Fig. 4.4).

The resource partitioning between both bird species is quite complete in the natural situation. Manipulation of the food stock, by planting big clams at a shallow depth, elicited interference between Curlew and Oystercatcher, which normally is very rare.

Male Curlews take a few clams only, whilst it is a main prey for the females. It is suggested that because males have a shorter bill than females, the proportion of accessible clams above the lower acceptable size limit is too small (Fig. 4.3, inset).

Oystercatchers deplete the clams in the winter following the second growing season. After this the Curlew females exert a heavy predation pressure on the remaining clams (Fig. 4.5).

Acknowledgements

The data summarized in this chapter are based upon the work done by many people. We would like to thank especially Piet Zegers for (among other things) catching the birds and Joke Bloksma, Bruno Ens, Peter Esselink, Rick Looijen, Bertrin Roukema, Marja de Vries and Renske de Vries who all participated as students (University of Groningen) in this part of the project. Without the help of Rudi Drent, Bruno Ens and Peter Evans this paper would not have been written. Jan Hulscher kindly permitted us to quote from his unpublished work. The figures were prepared by Dick Visser and the manuscript was typed by Mrs H. Lochorn-Hulsebos.

References

Beukema, J. J. (1979). Biomass and species richness of the macrobenthic animals living on a tidal flat area in the Dutch Wadden Sea: effects of a severe winter. *Neth. J. Sea Res.* **13**(2): 203–23.

Beukema, J. J., Bruin, W. de & Jansen, J. J. M. (1978). Biomass and species richness of the macrobenthic animals living on the tidal flats of the Dutch Wadden Sea: long-term changes during a period with mild winters. *Neth. J. Sea Res.* **12**(1): 58–77.

Brousseau, D. J. (1978). Population dynamics of the soft-shell clam *Mya arenaria*. *Mar. Biol.* **50**: 63–71.

Goss-Custard, J. D. (1980). Competition for food and interference among waders. *Ardea* **68**: 31–52.

Hulscher, J. B. (1982). The oystercatcher *Haematopus ostralegus* as a predator of the bivalve *Macoma balthica* in the Dutch Wadden Sea. *Ardea* **70**: 89–152.

Kenward, R. E. & Sibly, R. M. (1978). Woodpigeon feeding behaviour at brassica sites. A field and laboratory investigation of woodpigeon feeding behaviour during adaptation and maintenance of a brassica diet. *Anim. Behav.* **26**: 778–90.

Krebs, J. R. (1978). Optimal foraging: decision rules for predators. In *Behavioural ecology, an evolutionary approach*, ed. J. R. Krebs & N. B. Davies, pp. 23–63. Blackwell, Oxford.

MacArthur, R. H. & Pianka, E. R. (1966). On the optimal use of a patchy environment. *Am. Nat.* **100**: 603–9.

Myers, J. P., Williams, S. L. & Pitelka, F. A. (1980). An experimental analysis of prey availability for Sanderlings (Aves: Scolopacidae) feeding on sandy beach crustaceans. *Can. J. Zool.* **58**: 1564–74.

Reading, C. J. & McGrorty, S. (1978). Seasonal variations in the burying depth of *Macoma balthica* (L.) and its accessibility to wading birds. *Est. coast. mar. Sci.* **6**: 135–44.

Smidt, E. L. B. (1951). Animal production in the Danish Wadden Sea. *Meddr. Kommn. Danm. Fisk.-og Havunders.* **11**(6): 1–151.

Townshend, D. J. (1981). The importance of field feeding to the survival of wintering male and female Curlews *Numenius arquata* on the Tees estuary. In *Feeding and survival strategies of estuarine organisms*, ed. N. V. Jones & W. J. Wolff, pp. 261–73. Plenum Press, New York.

Vlas, J. de. (1979). Annual food intake by plaice and flounder in a tidal flat area in the Dutch Wadden Sea, with special reference to consumption of regenerating parts of macrobenthic prey. *Neth. J. Sea Res.* **13**(1): 117–53.

Zwarts, L. (1981). Habitat selection and competition in wading birds. In *Birds of the Wadden Sea*, ed. C. J. Smit & W. J. Wolff, pp. 271–9. Balkema, Rotterdam.

Zwarts, L. & Drent, R. H. (1981). Prey depletion and the regulation of predator density: oystercatchers (*Haematopus ostralegus*) feeding on mussels (*Mytilus edulis*). In *Feeding and survival strategies of estuarine organisms*, ed. N. V. Jones & W. J. Wolff, pp. 193–216. Plenum Press, New York.

5

Waterfowl movements in relation to food stocks

M. R. VAN EERDEN

Introduction

Many species of waterfowl that winter along the shores of western Europe breed at northern latitudes. In autumn they migrate towards the wintering areas which, according to the species concerned, coincide with, or lie south of, the zone free of snow and ice. In spring some species move north again, following the retreating ice edge (e.g. *Mergus* spp.), whereas others build up fat reserves at lower latitudes before migrating by a few long-distance flights to the breeding grounds (e.g. species of geese of the genus *Branta*). Radar studies have revealed the often massive flights of waterfowl that occur between staging areas (Bellrose & Sieh 1960; Bergman & Donner 1964; Blokpoel & Richardson 1978; Bergman 1978) and much attention has been paid to the modifying influences of weather conditions upon the timing of migration.

However, little is known of other influences on waterfowl movements and on the duration of use of possible wintering sites. We believe that the amount of available food at a given site plays an important role. For some arctic-nesting geese, the timing of spring migration normally coincides with the start of plant growth along a primarily south to north climatological gradient. Plants are most digestible during their first flush of growth, and, while on migration, the geese seem to ride the crests of the digestibility waves of their food plants (Drent, Weijand & Ebbinge 1978). In contrast, those waterfowl that are dependent upon highly mobile animal food may react to seasonal shifts in prey availability.

This chapter summarizes some of the work in progress at the Lauwerszee, northern Netherlands. This estuary borders the Dutch Wadden Sea and its enclosure by a dyke, built in 1969, caused marked changes in the food supply for waterfowl. The prime question we addressed was to what extent food supply influences bird numbers on a given staging area. We report here on the situation in autumn (August–December) as it refers mainly to three herbivorous species, Wigeon (*Anas penelope*), Barnacle Goose (*Branta leucopsis*) and Teal (*Anas crecca*). After some introductory remarks on each species we concentrate on the role of food depletion in affecting bird behaviour. If it is important, we would expect to observe, in parallel with changes in food, one or several of the following events (cf. Newton 1980): (i) shifts in bird numbers; (ii) shifts in diets; (iii) shifts in foraging speed; (iv) competition or related phenomena. We look first at food depletion in the area as a whole and then at the level of individual plants to assess which cues might be used by birds in deciding whether or not to feed there.

The study area and its vegetation

Following enclosure in May 1969, the Lauwerszee provided over 5000 ha of tidal sandflats where the vegetation was allowed to develop naturally; another 2000 ha of heavier soils were allocated to agricultural purposes and the original system of creeks and gullies made up the remaining 2000 ha. According to detailed studies by Joenje (1978), plant colonization started mainly from a seed bank on the former sea bottom. Within a few years the lowest parts of the vast sandflats were covered by an almost pure stand of glasswort (*Salicornia* spp.), at one time with an extent of about 3500 ha. Desalination of the fine-grained sandflats was slow and it required more than 10 years for plant succession to reach a stage no longer dominated by halophytes. At intermediate stages other species increased, at first members of the goosefoot family *Chenopodiaceae* (seablite (*Suaeda maritima*), sea purslane (*Halimione portulacoides*) and saltwort (*Salsola kali*)). Later these were invaded by sea spurrey (*Spergularia maritima*), sea aster (*Aster tripolium*), and grasses (*Puccinellia* spp. and *Agrostis* spp.). Because the area was not managed by man by grazing or mowing, each autumn a huge food stock for waterfowl was present in the form of seeds. After the growing season, the standing crop was not renewed and autumnal grazing by waterfowl could be studied rather easily.

Only a few food plants were common, *Salicornia europaea* being the most important, with a seed production peaking at over 340 tons of dry weight in 1972. Within one year after enclosure the approximately 2000 ha of water in the central part of the Lauwerszee became fresh. Sandflats suitable for wildfowl roosts and shallow freshwater areas for drinking remained available throughout the period of study (1969–81) and no hunting took place. So the main factor that varied between years was food abundance.

The methods of study of the birds and their foods

In the first six years detailed descriptions of plant colonization were made by Joenje (1978) and in the seventh by J. Prop & M. R. van Eerden. Assessment of total food stocks on the Lauwerszee required measurement of the extent, the density and the size of the food plants. After the first four years, changes in the total food stock of *Salicornia* mostly depended on shrinkage of the area covered. Annual mapping of vegetation was verified twice by aerial photography (1978, 1980, J. Slager & R. Drost, unpublished data).

More intensive measurements were carried out on a 44 ha tract of the area called Schildhoek. Grazing of seeds by wildfowl occurred predominantly at night and was measured directly by counting the seed heads lost, since the whitish stalks remaining at the base of the ear after grazing were very obvious. Because of the regular shape of the plants the number of seed heads grazed could be converted to estimates of the absolute amount of food removed.

Counts of droppings each morning indicated foraging patterns in the previous night. Droppings of *Anas penelope*, *A. crecca*, *A. acuta*, *A. platyrhynchos*, *Branta leucopsis* and *Anser anser* could be distinguished by their appearance. They are produced at a fairly constant rate and therefore indicate the time spent by a bird at a given spot. Thus when feeding on *Salicornia* Wigeon produce a dropping every 3 min and Barnacle Geese every 4.25 min on average.

Also, direct observations were made of foraging behaviour (time spent feeding, intake rate and flock density) from a 4 m tall tower hide during the daytime by telescope and at night by infra-red telescope. The 44 ha study area was divided into 1 ha cells, each watched continuously by a team of enthusiastic observers.

To calculate the number of bird-days spent by each species, 5–10 total bird counts were organized in the period August to December each year.

The annual cycles of the wildfowl species

The Barnacle Goose is a high-arctic breeder, spending four summer months there and the other parts of the year along the coasts of western Europe. Three distinct populations exist, the largest breeding in the Novaya Zemlya/Vaigach region and wintering in the Netherlands

(54 000 birds, Rooth *et al.* 1981). The migration route of the Russian population is well known through direct and radar observations. Fig. 5.1 depicts flyways and known staging areas *en route* (Bauer & Glutz von Blotzheim 1968; Kumari 1971; Owen 1980). On a European scale only a few places are known where flocks of more than one thousand birds can be seen and between which the birds distribute themselves in the winter.

Fig. 5.1. Migratory flyway of the Barnacle Goose (inset at top) with indication of principal autumn and winter haunts of the Novaya Zemlya population (centre) and location of Lauwerszee study area (inset bottom).

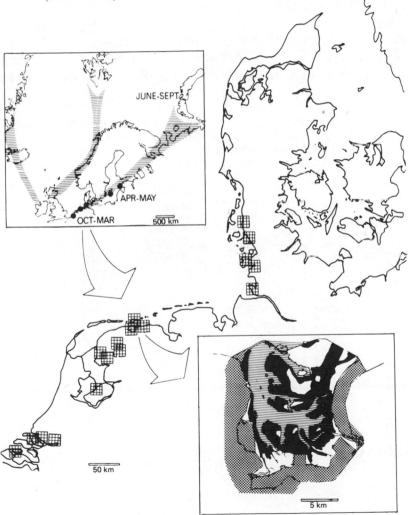

In a mild winter several thousands of birds tend to stay along the coast of northwest Germany (cf. Busche 1977); in a severe one almost all the geese concentrate into the Delta area of the Netherlands, the most southwestern part of the range. Our study area is one of the staging areas, shown in more detail in the bottom corner of Fig. 5.1. As many as 50 000 Barnacle Geese have been counted together at this site, representing over 95 % of the total flyway population.

A less discrete pattern of halting places is observed in Wigeon. As a grazing species, it is one of the most numerous wintering ducks in the Netherlands. Over 350 000 birds occurred in a series of mild winters (1974–78) (J. Rijnsdorp, unpublished). Some Wigeon feed on pastures, others on saltmarsh. The relative proportions of all Wigeon present in the Netherlands that feed in the two habitats change throughout the autumn and winter (Fig. 5.2). From September through to November most are found close to the sea in both the Wadden and Delta regions.

Fig. 5.2. Distributional shift of Wigeon wintering in the Netherlands (5-year means of regional counts). Inset shows habitat use through the season on the basis of bird-days (all areas combined) emphasizing the heavy usage of saltmarsh in autumn.

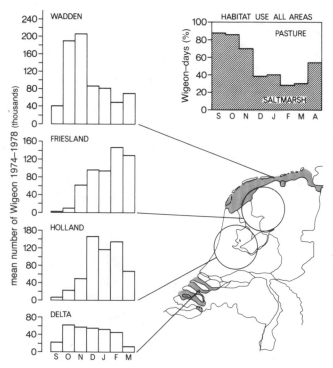

From December to March mainly inland foraging sites are used, chiefly in the provinces of Friesland, Noord Holland, Zuid Holland and Utrecht. Total numbers present in the Netherlands change rather little, however, suggesting a mass shift from salt- to freshwater plant foods. The use of the Lauwerszee area closely follows that of the Wadden Sea (Fig. 5.2) with up to 64 000 birds present in October. The major shift in habitat is related to a shift in the types of plants eaten, the saltmarsh plant species and algae being highly preferred in autumn.

Analysis of the diet (see Stewart 1967) of Wigeon in our study area also showed this shift, although both food types occurred in the same area. At the times of peak numbers in early autumn, *Salicornia* spp. is eaten in preference to the nearby grasses, whereas after most of the birds have gone the remaining Wigeon shift to a diet comprised almost totally of grasses. These findings could be explained if Wigeon obtain calorific returns from their foods similar to those measured for the *Branta* species of geese by Drent *et al.* (1978). They found the following values of energy assimilated (in kcal/100 g food taken): *Salicornia* 185 (October), *Enteromorpha* 184 (November), *Zostera* 160 (October), *Festuca* 157 (December), *Poa, Lolium* 142–158 (December–February). From a calorific point of view, birds should prefer to graze the foods highest on this energy gradient, if they are available. Apparently Wigeon do so. Whether it is really the number of calories or some other 'quality' factor in the food to which the birds respond is unknown but further investigation along these lines seems worthwhile. Clearly, colonization of vast areas of the Lauwerszee by *Salicornia* was of great importance to the birds. With the presence of large stocks of a highly preferred food the 'new' area competed easily with others on the flyway to attract birds.

The annual cycle and breeding areas of Teal largely overlap those of Wigeon (Bauer & Glutz von Blotzheim 1968). Most Teal that migrate to the North Sea area originate from the boreal zone of northwest Russia, west of the Urals (cf. ringing results, Speek 1975). These ducks use a highly specialized technique of sieving out food particles. The habitats they can use are thus restricted to those with a thin layer of water covering the feeding grounds. On migration their food consists mainly of vegetable matter. In western Europe no clearly defined halting places exist where feeding conditions are predictable and optimal year after year.

After September, large numbers appear suddenly soon after conditions become favourable at a given place, usually a result of flooding. The Lauwerzee area attracted up to 60 000 Teal in November 1974, the species being overall the third most common consumer of the seed stock present.

The impact of wildfowl on their food supplies, and vice versa

As mentioned above, one of the major food plants for waterfowl in the area during the years of our study was *Salicornia europea*. After germination in May from seeds lying in the bare mud, the next seed crop of this annual species ripens during October with a change in colour from green to deep red. Waterfowl begin to eat the crop soon after the period of ripening. Only plants bordering the water's edge and close to the roosts are consumed in the green stage.

Fig. 5.3 shows the relationship between standing stock and total food consumption by Wigeon and Barnacle Geese in the autumn of 1975. It is clear that the areas holding higher densities of food received higher grazing pressures (measured as biomass removed), but that the highest percentage of food was removed from the lower end of the food density gradient. A maximum of *c.* 70 % was grazed at biomasses of between 20–50 g dry weight/m², but only about 50 % at higher food densities. Each dot in the graph represents the value for a single area, the areas scattered over the total feeding grounds. So even at places where food was scarce, it was exploited by grazing Wigeon and Barnacle Geese provided that the plants formed an integral part of the vegetation sward of the flats (isolated patches of suitable plant foods were under-exploited).

Fig. 5.3. Grazing impact (dry weight of seeds removed/m²) in relation to food supply at the onset of autumn (biomass dry weight/m² of seeds on offer) for Wigeon feeding on *Salicornia* in the Lauwerszee.

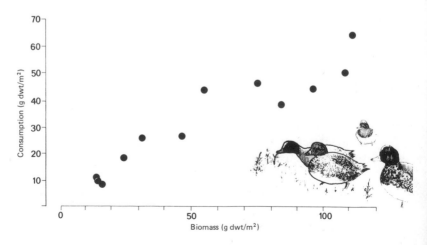

Compared to Wigeon, much lower proportions of seed stocks were eaten by Teal: 7 % of *Salicornia*, 8 % of *Spergularia* in 1975. This arises in part because the season during which Teal can exploit the seed stocks is short. Their specialized way of feeding is very much dependent upon suitable weather conditions (rainfall, inundation, and no ice). In addition we investigated whether Teal might require higher food densities for successful feeding than do other species, since, unlike their grazing allies, Teal filter individual seeds from a shallow layer of water rather than stripping whole seed heads.

We examined the grazing intensity of Teal on seeds of the grass *Agrostis stolonifera* by searching for droppings left in areas with a high

Fig. 5.4. Grazing of Teal in relation to food stocks in the Lauwerszee. Upper panel shows food on offer (see text for methods), lower panel grazing activity of Teal as revealed by droppings (note that grid is 50 × 50 cm).

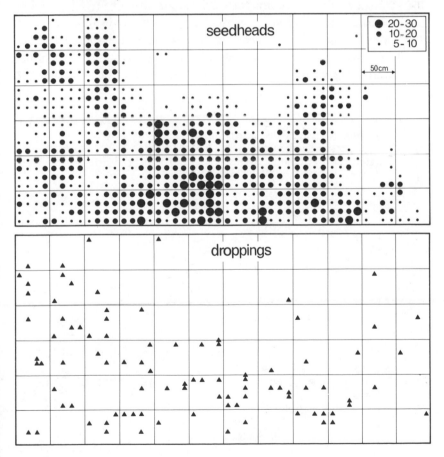

cover of vegetation, where floating plant leaves saved them from being washed away. Fig. 5.4 shows part of such a heavily vegetated area where the Teal used to forage at night. The patchy distribution of the food source is shown by the results of 1152 sampling quadrats and the cumulative location of the individual droppings is shown in the lower panel for comparison. Fig. 5.5 summarizes these results over a somewhat larger area and indicates the presence of a certain threshold level of food density, below which the birds did not respond to differences in the density of food present. Thus the small size of the seed particles ingested by Teal does seem to lead to the requirement of a high minimum seed density before exploitation occurs.

We then investigated the question of why not all the food potentially available is eaten by wildfowl, using Wigeon to examine in more detail the sequence of grazing events. By counting the droppings over the 44 ha study area we determined changes in foraging from night to night. Fig. 5.6 shows the total number of droppings found in 184 sampling quadrats (of $4 m^2$ each) which were cleared each day in the autumn of 1977 for 37 consecutive days. The pattern that emerged has been confirmed in other years and clearly shows that harvesting of seed crops by waterfowl is an intensive activity. It may be calculated that 500

Fig. 5.5. Grazing pressure (as measured by droppings deposited/m^2) in relation to food stocks (density of seedheads) for Teal feeding on *Agrostis* seed in the Lauwerszee during October–November 1980 (constructed from Fig. 5.4).

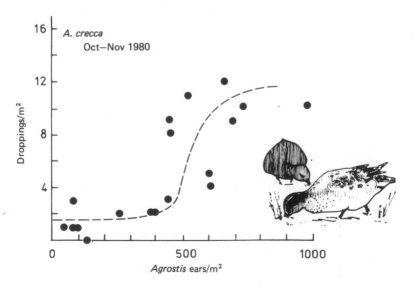

droppings found correspond to about 4500 Wigeon foraging continuously for 10 hours at night. During a total of only 11 nights, 80 % of the total annual grazing pressure was exerted.

The most important features illustrated in Fig. 5.6 are the discontinuities in exploitation of the seed crops. Grazing periods of several successive nights alternated with periods of little grazing, when the huge flocks of birds were feeding on neighbouring areas. This regular shifting of feeding grounds resulted in a gradual lowering of the food stocks over the whole area. Associated with this, the diet of the birds changed. The proportion of *Salicornia* dropped from 90 % in the first two grazing periods to only 30 % in the last period, when it was replaced chiefly by perennial grasses. During the first grazing period the birds removed only 10–20 % of the total amount of *Salicornia* seeds available to them. These almost always proved to be only the uppermost ears of each plant. Subsequent grazing resulted in the harvest of the remaining (lower) parts of the plants, until between 50 and 70 % of the initial stock

Fig. 5.6. Grazing rhythmicity in Wigeon utilizing pioneer vegetation in the Lauwerszee. Three waves can be distinguished, and as the seed stock was depleted a shift in diet occurred (see bar graphs).

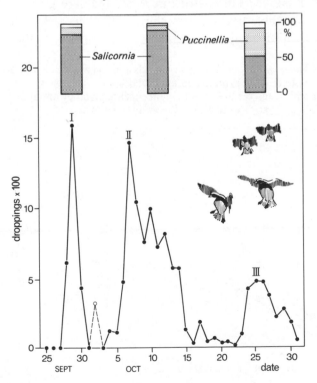

was removed, as mentioned earlier. Substantial differences in weight and protein content exist amongst the ears of *Salicornia* occurring at different heights above the ground (Fig. 5.7). In a sample of plants of 12 to 15 cms in height, there was a sixfold increase in mean dry weight from the lowest to the topmost ears and a threefold increase in weight of protein and in energy content. These quality gradients within single plants explain why Wigeon strongly prefer to graze the uppermost ears. They also imply that, if grazing at a constant speed, a foraging bird would achieve a progressively lower rate of intake as the total remaining food stocks decline. We might expect them to compensate for this by increasing peck frequency, a phenomenon well known in geese grazing on progressively shorter grass swards during winter (Owen 1972).

Nocturnal observations on foraging Wigeon, however, showed in fact that they decreased the number of bites made per minute. If we assume that each peck resulted in the removal of a single ear (as seems probable because of the spacing of the ears on the plant), the birds did not compensate for the lowering of intake rate. Despite the higher densities of ears lower down the plants, the ducks apparently are hindered while foraging by the increasing density of empty stalks. Except by trampling, they do not alter the vegetation structure while they are grazing but merely remove the edible fraction, as if they were giants picking fruits from an orchard whilst standing above the tree canopy. Since they do

Fig. 5.7. *Salicornia* seedhead quality in relation to height of ear above ground. Note that the protein content (solid squares) shows a maximum plateau. Weight of individual ears are given by open squares. Sample sizes were above 50 except as shown.

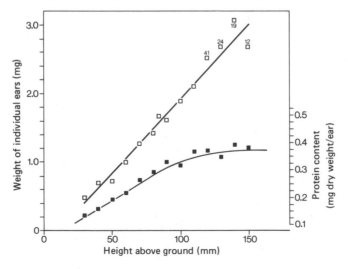

not feed faster as they graze down the plants, they must feed for a longer period if they are to balance their daily energy expenditure (DEE) later in the autumn. We have calculated that the birds taking part in the first grazing period (Fig. 5.6) could balance their DEE by less than five hours foraging per night. During the second period the time needed should be only seven hours, but in the third period they would have needed 20 hours of foraging on *Salicornia* alone. In fact, the ducks then shifted their diet towards the more abundant but less preferred grasses, with only one third of their food intake still being *Salicornia* (Fig. 5.6). Since Wigeon are completing their postnuptial moult in October and may be laying down fat reserves for the winter (Bauer & Glutz von Blotzheim 1968) it is not surprising that they should change their diet to increase their rate of intake. Apparently fewer birds were involved in the last grazing period and at the same time many Wigeon left the Lauwerszee.

The combined effects of grazing by several wildfowl species
So far we have discussed several aspects of the complex relationship between birds and their food, for each species separately, and the question now arises of whether competition between species might play a role when several tens of thousands of birds are present together. The species most likely to compete with each other for food are Barnacle Goose and Wigeon, both grazing species. As noted earlier, Wigeon tend to arrive in autumn at least one month before Barnacle Geese (Wadden Sea area, Joensen 1974; Busche 1977; P. Zegers, personal communication). When the geese arrived at the Lauwerszee, the *Salicornia* sward was already being harvested by Wigeon. The geese foraged by day and thereby alternated use of the area with the ducks, which fed at night. Often they foraged in places overlooked by the ducks. In such places, the intact sward still showed as a reddish purple colour, easily spotted from the air by the geese during daylight.

Extensive counts of droppings of both species on our study areas supported the idea of complementary feeding by the ducks and geese. Fig. 5.8 shows the relative grazing pressure of both species in a number of sites. Those spots heavily grazed by Wigeon received little attention from geese, and vice versa. The inset in Fig. 5.8 shows the total impact on the vegetation of the whole Lauwerszee by the grazing birds in 1975. Wigeon consumed 46%, Barnacle Geese 24% and the highly specialized Teal only 14% of the total seed crop of 200 tons dry weight of *Salicornia*. In other years this pattern was repeated, Wigeon always taking the greater part of the annual harvest. In Fig. 5.9 the total food consumption by grazing birds is shown to follow a pattern set by annual changes in seed stocks, caused by plant succession. In 1971 the mudflats were totally covered by *Salicornia* for the first time (Joenje 1978). In that year Barnacle Geese managed to eat a greater proportion of the seeds than usual since many of the plants were of the tall (*stricta*) type, partly out of reach of Wigeon. The next year, 1972, plant height was considerably lower (*brachystachia* type); total production reached a peak value, and so did total consumption, that by Wigeon by far exceeding that by Barnacle Geese. From 1972 onwards, the Barnacle Geese gradually advanced their date of first arrival in the area, by about three weeks over a period of eight years. As a result they took an increasing proportion of the total seed harvest (Fig. 5.9) removed by the two species. In 1979 this trend was broken but in 1980 the geese again arrived early. However, they were not 'rewarded' in this year by gaining more food (because total stocks were lower) and we expect in the years to come that geese will no longer overfly other possible halting places on their way south as they did in the 1970s.

Fig. 5.8. Barnacle Goose (*Branta leucopsis*) visitation (y axis, as measured by droppings deposited/m²) in relation to prior usage by Wigeon of the same *Salicornia* vegetation (x axis, again in droppings/m²). Inset shows proportion of seed crop of *Salicornia* removed by wildfowl in the 1979 season (23 % ungrazed remaining) and arrival and departure dates for the study area (black bars indicate mean *Salicornia* feeding periods).

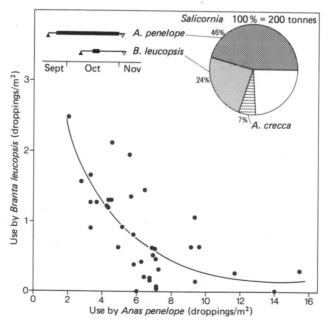

The Lauwerszee provided huge amounts of food for waterfowl during the early plant succession. Now, 12 years after enclosure, the *Salicornia* swards have been replaced by perennial grasses. In autumn the seed heads of *Agrostis stolonifera* provided food for Greylag Geese (*Anser anser*). Again the birds are highly selective, stripping the seed heads but leaving the grass blades untouched. In about four weeks during September, 3500 geese are able to harvest up to 75 % of the total seed crop on offer. Experiments with captive Greylags in small enclosures in the field are underway to see whether the observed sequence of food plants

Fig. 5.9. *Salicornia* seed consumption by Wigeon and Barnacle Goose (metric tonnes dry weight) in relation to total crop in the Lauwerszee following closure in 1969. Open circles show the arrival date (in October) of geese in each year.

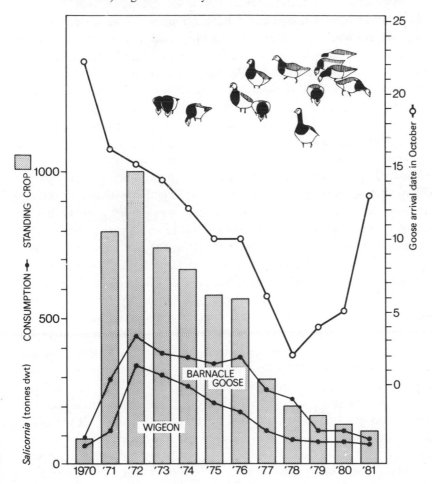

taken can be explained in terms of decreasing profitability. Preliminary results indicate that, by seed stripping, geese might reduce the time needed for foraging to about half that required for foraging on the less preferred grass blades. Indeed, many of the wild geese leave the area and fly to their Iberian wintering sites once the seed stocks have been depleted on the Lauwerszee.

During the 1970s the shallow fresh water of the former Lauwerszee has been colonized by macrophytes, predominantly the pondweed species *Potamogeton pectinatus*. The leaves and seeds of this submerged vegetation are consumed by Coots (*Fulica atra*), Wigeon and Gadwall (*Anas strepera*). The numbers of these species present in late summer vary roughly in proportion to the extent of the areas with a high surface cover of *Potamogeton*. Following this period of exploitation, the floating plant material dies off and another consumer, Bewick's Swan (*Cygnus columbianus bewickii*) arrives. *Potamogeton* tubers lie hidden in the mud but are within reach of the trampling and digging swans even in water depths of as great as 1 m. As in the case of Teal, the swans cannot see their food but have to locate the better feeding areas by touch. Detailed observations at night, when most feeding occurred, showed that the swans did indeed concentrate in areas of the highest tuber biomass. Once again, a relatively short period of harvest was observed, lasting only four weeks from mid-October, by up to 700 birds. An apparent threshold density for tuber exploitation was determined as being *c*. 7 g dry weight/m^2, and in areas above this density level, 43 % of the tuber stock was removed by the swans. Together with other avenues of loss, almost 70 % of the tuber biomass was depleted in the better areas, resulting in a low and almost uniform tuber density at the close of exploitation. At this point many swans leave the area, and the remainder shift to field-feeding where remnants of sugar beet and potatoes, left behind after the harvest, can be found. A calculation of the daily ration (tubers disappearing from the study plot, corrected for non-grazing losses and divided by the number of swan-nights determined by direct observation) gave an estimate of 283 g dry weight per swan-night, in reasonably close agreement with McKelvey's (1980) estimate for the Trumpeter Swan of 332 g dry weight (the body weights of the two species are respectively 6050 g and 11 400 g).

One of the main conclusions regarding the effects of vegetation density on grazing waterfowl must be that although not all the plant food available was eaten, when translated into an amount of energy that the birds could extract profitably, probably the upper limit was reached in most years. The number of bird-days of use of a given area therefore largely depends on the amount of food available in that area, but is also influenced by that in other areas along the migration flyway. The use made of a given area may therefore be set both by internal (food within the area) as well as by external factors (food between areas, specific needs other than energy and protein).

Acknowledgements

Many persons were involved in the various aspects of the work. Tj. Reitsma of the Ijsselmeerpolders Development Authority kindly gave permission to work in the area for many years. S. Keuning, J. Onderdijk, E. Schuldink and W. de Vries of the RIJP were always willing to solve many technical problems that arose during the fieldwork. J. Koenes from the Zoological Laboratory, Haren, provided good waterproof tower hides and kept a dinghy going. As a part of their study, M. Boeken, J. Beekman, J. Bloksma B. Ens, H. van Huffelen, G. Jaspers, D. Prop, G. Smits and R. Ydenberg took part in the project. Their enthusiasm guaranteed that the pile of data grew quickly. The long-term counting of birds would not have been possible without the help of a group of keen observers willing to take part in bird counts often announced at short notice. R. Drent, J. Hulscher, D. and J. Prop, J. Tinbergen and J. Veen proved to be the most tough of these throughout the years.

J. Bottenberg provided chemical analysis of food plants, D. Visser typed the manuscript.

The often painstaking gathering of data would not have been possible without the belief that this type of fieldwork might lead to a worthwhile harvest from nature. In a stimulating way Dr R. H. Drent guided me very much towards this belief. A large part of the study was undertaken from the University of Groningen and finally I'd like to acknowledge greatly the companionship of Jouke Prop, with whom I spent much time sorting out plants, bird data and my sometimes wind-blown mind.

References

Bauer, K. M. & Glutz von Blotzheim, U. N. (1968). *Handbuch der Vögel Mitteleuropas*, Band 2. Akademische Verlagsgesellschaft, Frankfurt am Main. 528 pp.

Bellrose, F. C. & Sieh, J. G. (1960). Massed waterfowl flights in the Mississippi Flyway, 1956 and 1957. *Wilson Bull.* **72:** 29–59.

Bergman, G. (1978). Effects of wind conditions on the autumn migration of waterfowl between the White Sea area and the Baltic region. *Oikos* **30:** 393–7.

Bergman, G. & Donner, K. O. (1964). An analysis of the spring migration of the Common Scoter and the Long-tailed Duck in southern Finland. *Acta Zool. Fenn.* **105:** 1–59.

Blokpoel, H. & Richardson, W. J. (1978). Weather and spring migration of Snow Geese across Manitoba. *Oikos* **30:** 350–63.

Busche, G. (1977). Zum Wintervorkommen der Nonnengans (*Branta leucopsis*) an der Westküste Schleswig–Holsteins. *Vogelwarte* **29:** 116–22.

Drent, R., Weijand, B. & Ebbinge, B. (1978). Balancing the energy budgets of arctic breeding geese throughout the annual cycle: a progress report. *Verh. orn. Ges. Bayern* **23:** 239–64.

Joenje, W. (1978). Plant colonization and succession on embanked sandflats. Dissertation, University of Groningen, Netherlands.

Joensen, A. H. (1974). Waterfowl populations in Denmark 1965–1973. *Dan. Rev. Game Biol.* **9:** 1–206.

Kumari, E. (1971). Passage of the Barnacle Goose (*Branta leucopsis*) through the Baltic area. *Wildfowl* **22:** 35–43.

McKelvey, B. (1980). Studies of Trumpeter Swans wintering in British Columbia. Unpublished PhD thesis, Simon Fraser University, Burnaby, Canada.

Newton, I. (1980). The role of food in limiting bird numbers. *Ardea* **68:** 11–30.

Owen, M. (1972). Some factors affecting food intake and selection in White-fronted Geese. *J. anim. Ecol.* **41:** 72–92.

Owen, M. (1980). *Wild geese of the world.* B. T. Batsford Ltd, London. 236 pp.

Rooth, J., Ebbinge, B., van Haperen, A., Lok, M., Timmerman, A., Philippona, J. & van den Berg, L. (1981). Numbers and distributions of wild geese in the Netherlands 1974–1979. *Wildfowl* **32:** 146–55.

Speek, B. J. (1975). Bird ringing in the Netherlands. *Limosa* **48:** 129–57.

Stewart, D. R. M. (1967). Analysis of plant epidermis in faeces: a technique for studying the food preferences of grazing herbivores. *J. appl. Ecol.* **4:** 83–111.

6

Diving duck populations in relation to their food supplies

O. PEHRSSON

During the 10-year period from 1964 to 1975, wildfowl counts were made during the non-breeding season along the greater part of the Swedish west coast (Fig. 6.1). For some species, food selection and food abundance were also investigated (Pehrsson 1973, 1976a,b,c, 1978). After some years it was found that the numbers of certain northern breeding duck species fluctuated from year to year in a similar way, but that the pattern of annual change in more southerly breeding species differed from this. Accordingly, studies were made during a six-year period in a breeding area in northernmost Sweden (Fig. 6.1). Also, the regional occurrence of breeding waterfowl in the archipelago of the west coast of Sweden was mapped as a result of a survey of about 2000 islands and skerries.

The primary aim of these investigations was to document annual fluctuations in duck numbers and, if possible, to establish their causes. It was thought that factors regulating or limiting duck populations might be discovered through natural experiments in time and space, if studies were continued for a sufficient length of time.

This chapter summarizes published and unpublished results, indicating how abiotic factors affect important food organisms for waterfowl on the coast, and how groups of herbivorous and piscivorous species (Table 6.1) and, in particular, individual diving duck species are dispersed in winter in relation to their food stocks.

Methods used and presentation of results
Non-breeding birds were counted along the Swedish west coast, from land-based observation points, up to nine times each year between August and April (Pehrsson 1975). Three times each year, from

Table 6.1. *Groups of herbivorous birds and fish feeders*

Herbivorous birds	Fish-feeders
Mallard (*Anas platyrhynchos*)	Divers (*Gavia* spp.)
Teal (*Anas crecca*)	Grebes (*Podiceps* spp.)
Wigeon (*Anas penelope*)	Cormorant (*Phalacrocorax carbo*)
Pintail (*Anas acuta*)	Heron (*Ardea cinerea*)
Mute Swan (*Cygnus olor*)	Red-breasted Merganser (*Mergus serrator*)
Whooper Swan (*Cygnus cygnus*)	Goosander (*Mergus merganser*)
Coot (*Fulica atra*)	Razorbill (*Alca torda*)/
	Guillemot (*Uria aalge*)
	Black Guillemot (*Uria grylle*)
	Harbour seal (*Phoca vitulina*)

Fig. 6.1. The study area on the Swedish west coast (rectangle) and in northernmost Sweden (filled circle). Inset: the Swedish west coast. A–N = areas mentioned in the text, RN = the River Nordre älv, RG = the River Göta älv.

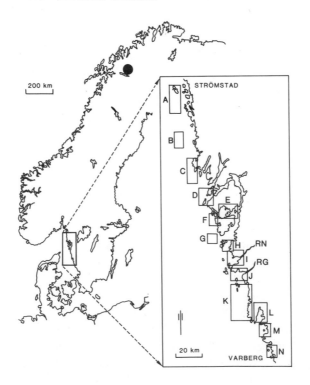

November to March, these surveys were complemented with counts from coastguard boats in the outer archipelago. Most results are summarized in map form. A few diagrams include actual counts, one uses only maximum counts, but most indicate the relative importance of different sites, using an index described in detail by Pehrsson (1979). This index is derived from the maximum count in a particular year, at the locality chosen, expressed as a proportion of the sum of the maximum counts at all localities in that year. This proportion is then averaged over the number of years counted. (For evaluation of the relative importance of sites for whole groups of species, the indices for all individual species were summed for that site.)

General features determining feeding conditions

The utilization by waterfowl of localities along the Swedish west coast may be examined in relation to gradients of nutrient supply, exposure to the sea or to the wind, and the strength of estuarine and tidal water circulation. (Closeness to a migration route may also be important for some waterfowl species and will be considered later.) The physicochemical factors affect both the quantity of foods present and their availability to the birds in the following ways, some of which changed in importance during the study period.

Nutrient supply

Most nutrients reach coastal waters via rivers. On the Swedish west coast, the largest freshwater inflows come through the Rivers Nordre älv ($450\,m^3/s$) and Göta älv ($150\,m^3/s$) (RN and RG flowing into estuaries I and J, in Fig. 6.1). The coastal Baltic current brings nutrients northwards along the coast. The nutrient content attributable to each freshwater outlet decreases with increasing distance from that outlet (Eriksson, Peippe & Ryberg 1980), while the nutrient content of the coastal current varies along the coast, depending on losses to primary production and on addition of nutrients from more outlets.

The type of nutrients brought to the estuaries in our study areas changed considerably during the 10-year study. During the 1960s, sewage from urban areas was discharged into the River Göta älv without any treatment. Consequently, at this time most nutrients were locked in an organic form and larger particles were utilized directly by detritus-feeding organisms in the *Macoma* community (Fig. 6.2). Smaller particles were utilized by the blue mussel (*Mytilus edulis*) further out to sea in the archipelago. However, because too much material was discharged too quickly, the benthic community became extinct in parts of the estuary (Fig. 6.2). Fluctuations in the environmental conditions, leading to the death of some sedentary animals, periodically favoured scavengers like the estuarine amphipods (*Gammarus* spp.) (Pehrsson 1976*b*).

Since the early 1970s most of the sewage from the Göteborg area has been treated before discharge into the Göta älv estuary (J in Fig. 6.1). Thus the supply of organic detritus to both estuaries has decreased considerably, and fauna had returned to the 'dead' areas of the Nordre älv estuary (I in Fig. 6.1) by 1972 (Fig. 6.2). However, the supply of nutrients, rather than detritus, to the estuaries and archipelago has not decreased to the same extent, since many nutrients are now brought to the river in inorganic form. For example, almost 60% of the total phosphorus from the Göteborg area is still added to the Göta älv estuary. Inorganic forms of nutrients probably allow clearer water to persist in the estuaries, but they favour primary production, both of macrophytes in shallow estuarine surface waters and of phytoplankton further out in the archipelago, where turbidity decreases even further with distance from the river outlet. Blue mussel production may, in turn, be favoured at these larger distances from the estuaries. Because the freshwater outflow currents turn to the right in northern hemisphere estuaries, the northern side of the Swedish west coast estuaries obtain

Fig. 6.2. Benthic sampling in the Nordre älv estuary (Fig. 6.1, I). Living animals were present in the whole of the 1972 sampling area (from Pehrsson 1976*b*).

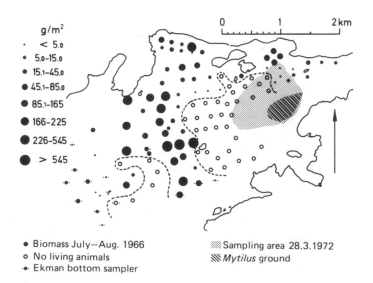

g/m^2

·   < 5.0
•   5.0–15.0
●   15.1–45.0
●   45.1–85.0
●   85.1–165
●   166–225
●   226–545
●   > 545

• Biomass July–Aug. 1966
o No living animals
-•- Ekman bottom sampler

▨ Sampling area 28.3.1972
▧ *Mytilus* ground

the highest nutrient supply, and the archipelago northwards of each estuary is influenced also (Pehrsson 1976*a*). Moreover, the nutrient content of the northward-flowing coastal current is enriched, as mentioned earlier.

Effects of exposure and water circulation

The dominant wind in this part of northwestern Europe blows from the southwest. Such winds cause an almost perpetual sea swell against the submerged exposed skerries along the Skagerak (northern) part of the Swedish west coast. On the Kattegat (southern) part of the coast, however, the sea swell is of much shorter duration after strong winds, because this stretch of sea is sheltered to the west by Denmark. Water turbulence caused by the swell is favourable to the growth both of blue mussel populations and of crustaceans such as *Idotea* spp. that live on these rocks, because they receive more food items per unit time than in still water.

The tidal amplitude is low on the west coast of Sweden (less than 35 cm) but this, in combination with meteorological changes affecting air pressure, is sufficient to create tidal currents inshore from the larger islands and in the mouths of fjords. These currents provide the most rapid growth in mussel beds anywhere along the coast.

Shallow, exposed fjords and bays in this part of Sweden tend to have sandy bottoms because clay particles are carried to sheltered or deeper areas by the extensive water circulation. This results in more turbid water and lower primary production than in shallow sheltered sites with clay bottoms, where submarine meadows of *Ruppia*, *Zostera marina*, *Potamogeton pectinatus* and *Chara* occur. Only in nutrient-enriched sites are exposed sandy bottoms colonized, by *Ulva lactuca*, which can attach to a hard surface.

Water circulation in an estuary results not only from exposure to the sea, but also from inflowing fresh water. This produces a surface current which turns to the right in the northern hemisphere. On the opposite (left) side of the estuary a compensation current of salt water enters, flows along the bottom and eventually mingles with the fresh water. This clear saline water inflow may ensure the survival of a good benthic community even in a heavily polluted river mouth (but only on the left side) (Pehrsson 1976*a*). Also, it may provide good foraging conditions near the bottom for fish-eating birds, which can dive through the turbid surface fresh water. This is particularly important in the Nordre älv estuary, from which three times as much fresh water is discharged as from the Göta älv and where, therefore, the compensation current is expected to be three times as great.

The winter distribution of coastal herbivorous waterfowl

Herbivorous birds, considered as a group (Table 6.1), are often concentrated near estuaries but also occur in sheltered, unpolluted areas which have muddy bottoms and submarine meadows (Fig. 6.3a). In the southern part of the west coast of Sweden, estuarine conditions near sewage treatment plants were used most. For example, in Kungsbackaf-jorden (Fig. 6.1, L), use decreased with increasing distance from the freshwater outflow. Herbivorous birds were more numerous in the Göta

Fig. 6.3. Distribution of herbivorous waterfowl (*a*) and fish-feeding birds (*b*) on the Swedish west coast. Symbol size proportional to relative importance of sites.

älv estuary than in the Nordre älv estuary, where the turbid river water impeded primary production. A richer inorganic supply and less water circulation in the Göta älv estuary may have favoured growth of *Ulva lactuca*, which is the most important plant food in the estuary. Very few herbivorous birds utilized the exposed Vendelsöfjord (Fig. 6.1, M), because of the low macrophytic production there.

The winter distribution of coastal fish-feeding birds

Fish-feeding birds (Fig. 6.3*b*) show a distribution pattern almost complementary to that of the herbivorous birds. For example, in Kungsbackafjorden (and in other areas) fish-feeders increased with increasing distance from the freshwater outflow. Even in the two estuaries of Göta and Nordre älv this pattern seems to hold.

Fish-feeding birds avoid turbid estuarine water, but further from the river outlets, where water is clearer, they (as well as the fish) obviously respond to higher levels of secondary production. They are more abundant south of the Göteborg area (Fig. 6.1, K) than outside and to the north of the large Göta and Nordre älv estuaries. This difference may be ascribed to the effects of turbid river water entering the coastal Baltic current and passing northwards. As already mentioned, in the Nordre älv estuary a vertical turbidity gradient exists, with clear saline water at the bottom forming the compensation current. In this area the fish-feeders may utilize both the clear water within the estuary and at localities some distance away from it, where turbidity also decreases.

The distribution and numbers of diving ducks in relation to their foods

Most diving ducks feed on benthos. The Eider Duck (*Somateria mollissima*) is a common breeder along the Swedish west coast, with the most dense breeding populations outside the river estuaries. The most frequented breeding and nursery areas are those influenced chiefly by the nutrient-rich river waters which encourage growth of the blue mussel, the main food of the Eiders (Pehrsson 1976*a*). In winter most Eiders leave the coast and move to Danish waters, returning by mid-March (Pehrsson 1978). In winter and until March, birds are found in the outer Skagerak archipelago, where they feed on the blue mussel beds on the exposed skerries. Here the ducks must improve their nutritional condition before returning to the inner archipelago to lay eggs, starting at the beginning of April.

During the winters of 1969 and 1970, ice formed even on the sea and the most exposed mussel beds were destroyed by mechanical grinding by the pack ice (Fig. 6.4). The ducks moved further into the archipelago to search for food on less-rich feeding grounds never utilized in other years. Associated with these shortages of winter food were appreciable declines in the breeding population. During the 1970s the population has increased considerably (Åhlund 1980), possibly through the effect on the mussel stocks of increased nutrient supply from sewage treatment plants and from leaching into the rivers of agricultural and forest fertilizers.

Wintering Tufted Ducks (*Aythya fuligula*) are found almost exclusively in the river estuaries (Fig. 6.5). In autumn and spring they share their feeding grounds with migrating Goldeneyes (*Bucephala clangula*) (Fig. 6.6). These are found chiefly along the central part of the coast. During severe conditions all Tufted Ducks are confined to the Göta älv estuary,

Fig. 6.4. Numbers of Eiders on the Swedish west coast in relation to winter temperature and availability of blue mussels (*Mytilus edulis*).

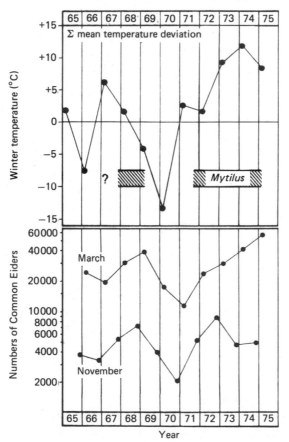

where the shipping traffic creates open water and wave-exposed blue mussel feeding grounds. The number of wintering Tufted Ducks has increased in parallel with the breeding Eider population which uses the same estuarine areas as nursery grounds (Fig. 6.7).

Whereas the Eider Duck utilizes blue mussel beds wherever these occur (Pehrsson 1978), the Tufted Duck uses only those in the estuaries in mid-winter, while at the same time of year the Long-tailed Duck

Fig. 6.5. Distribution of Tufted Ducks (*Aythya fuligula*) along the Swedish west coast. Symbol size proportional to the relative importance of sites.

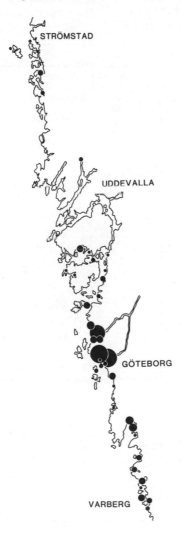

(*Clangula hyemalis*) is found only at the most exposed skerries (Fig. 6.1, A, B, C and G) along the Skagerak coast (Fig. 6.8). Only small numbers of Long-tailed Ducks are found on the Swedish west coast in comparison with numbers in the Baltic, where they forage entirely on mussels. On the west coast of Sweden, however, they take chiefly the crustacean isopod *Idotea pelagica*, as found by stomach analyses (O. Pehrsson & S. Persson, unpublished). When the mussel-feeding Eiders left the exposed

Fig. 6.6. (*a*) Occurrence of Goldeneyes (*Bucephala clangula*) along the Swedish west coast in October and March–April. Symbol size proportional to the maximum number of birds present at each site during the years indicated. (*b*) Migration corridors of this species through Scandinavia (from Pehrsson 1975).

(a)

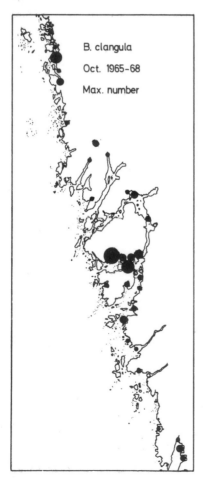 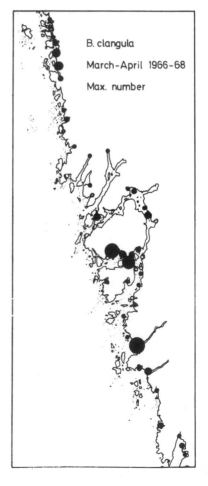

(b)

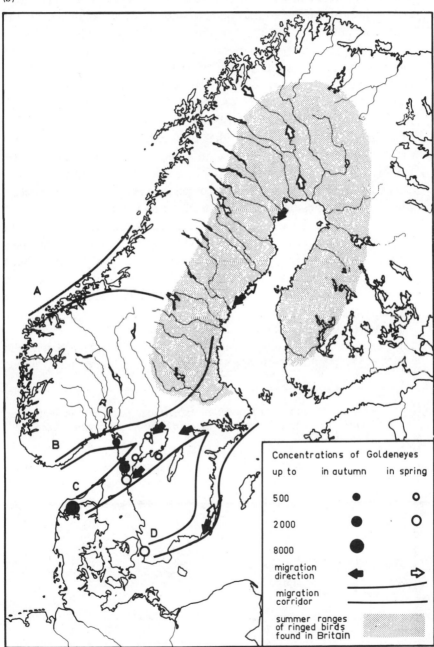

Concentrations of Goldeneyes

up to	in autumn	in spring
500	●	○
2000	●	○
8000	●	○

migration direction ◀ ⇨

migration corridor

summer ranges of ringed birds found in Britain

skerries in the icy conditions of 1969 and 1970 (because of the destruction of the blue mussels) the crustacean-feeding Long-tailed Ducks remained there, because their prey were not affected by the ice. The results presented above indicate that the Eider population breeding on the Swedish west coast may be determined by food abundance in the wintering areas, before the breeding season starts. Similarly, after severe winters the breeding populations of Tufted Duck in southern Sweden and on the Swedish west coast have decreased, although the wintering populations have increased both in Sweden (Fig. 6.7 and Nilsson 1978) and in England (Atkinson-Willes 1963). These observations suggest that populations of Eider, Tufted Duck and probably also Mute Swan (*Cygnus olor*) and other birds are regulated in their winter range. The reason for the increase in numbers of these species may be the increased supply of nutrients to their coastal winter feeding grounds.

The Long-tailed Duck, however, belongs to another group of species whose populations are believed to be regulated on the breeding grounds. The chief foods of the Long-tailed Ducklings are the crustacean *Polyartemia forcipata* and other euphyllopods, found only in lakes without fish (Pehrsson 1973, 1974). In recent centuries fish have been introduced by man into many northern Scandinavian lakes (Ekman 1910), which must have caused a considerable decline in suitable breeding areas for this duck species.

In the years when small rodents are abundant, predation pressures on Long-tailed Ducks are reduced and twice as many ducklings hatch as in other years (Pehrsson 1976*c*). But during peak years for rodents,

Fig. 6.7. Number of Eiders in the estuarine nursery areas in September and Tufted Ducks in the same areas in October–December (from Pehrsson 1976*a*).

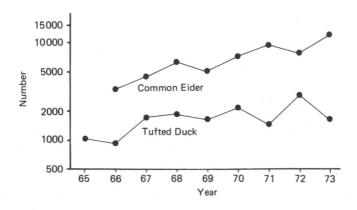

ducklings have to travel twice as much as usual between lakes to find food, and small ducklings grow more slowly than in other years. These findings indicate that the Long-tailed Duck population approaches the maximum possible level during a rodent peak year. Changes in the abundance of food in the winter are thus unlikely to affect the population levels.

Fig. 6.8. Distribution of Long-tailed Ducks (*Clangula hyemalis*). Symbol size proportional to maximum numbers present at each site in January.

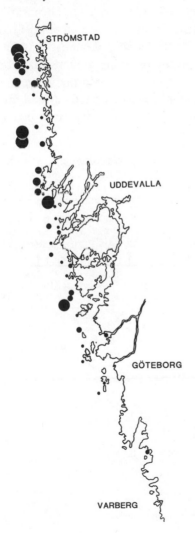

Partitioning of a food resource amongst different duck species

The blue mussel is utilized by several diving duck species, which could lead to interspecific competition. The maximum length of mussels consumed by species with different weights is shown in Fig. 6.9, which suggests that use of mussels may be partitioned by size. The larger Eider (Sm) and Common Scoter (*Melanitta nigra*) (Mn) may leave the Baltic in winter because the mussels there are large enough to satisfy only Long-tailed Ducks (Ch) and Tufted Ducks (Af), which are both lighter and proportionately smaller in size (Pehrsson 1976*b*). The Goldeneyes (Bc) feeding during migration, together with Tufted Ducks, in the west coast estuaries select particularly small food items, perhaps in order to avoid competition. The mussel resource may even be partitioned by habitat, for instance in the Baltic, where the similar-sized Long-tailed and Tufted Ducks utilize different feeding grounds. But in species whose numbers are regulated on the breeding grounds, like the Long-tailed Duck, competition is less likely to occur in the winter range.

Increased mussel production implies more abundant food for several diving duck species, but not all may be expected to react by increasing

Fig. 6.9. Maximum length of blue mussels consumed by Goldeneyes (Bc), Velvet Scoters (*Melanitta fusca*) (Mf), Long-tailed Ducks (Ch), Tufted Ducks (Af), Scaup (*Aythya marila*) (Am), Common Scoter (*Melanitta nigra*) (Mn), and Eider (Sm) in relation to the weight of the ducks (from Pehrsson 1976*b*).

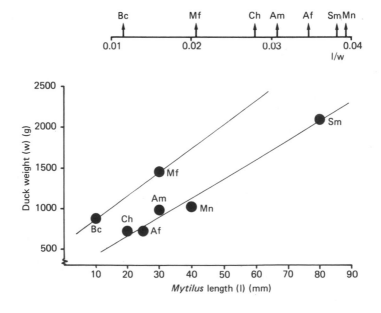

their populations, as mentioned earlier. Thus efforts at management of the winter areas are not likely to benefit populations that are regulated on the breeding grounds. Neither will management of the breeding grounds favour populations that are regulated in the winter areas.

Consequently, the first step in management research must be to establish where a given species is regulated. There is thus a need to make investigations both in the breeding and wintering areas, preferably in the same years. Comparative studies of both poor and rich breeding (and wintering) habitats are desirable. The condition of both breeding and wintering birds is probably the most important criterion to be measured, and careful study of the factors behind fluctuations in numbers between years may give valuable information about limiting and regulating mechanisms.

Acknowledgements

I wish to thank a large number of wildfowl counters who assisted during the years of this study, and Professor Ralph Smith for critically reading an earlier draft.

References

Åhlund, M. (1980). *Förändringar i häckfågelfaunan på ett antal fredade och ej fredade öar i Bohuslän mellan 1966 och 1979*. Länsstyrelsen i Göteborgs och Bohus län. 52 pp.

Atkinson-Willes, G. L. (ed.) (1963). *Wildfowl in Great Britain*. Monographs of the Nature Conservancy No. 3, London. 368 pp.

Ekman, S. (1910). Om människans andel i fiskfaunans spridning till det inre Norrlands vatten. *Ymer* **30**: 133–40.

Eriksson, E., Peippe, J. & Ryberg, E. (eds.) (1980). *Recipientundersökningar år 1979–1980. Göteborgs-regionens Ryaverk AB:s Vattenvårdsanläggningar*, Rapport 14. 90 pp.

Nilsson, L. (1978). Internationella sjöfågelinventeringarna i Sverige 1972/73–1975/76. *Vår Fågelvärld* **37**: 19–32.

Pehrsson, O. (1973). Chief prey as a factor regulating populations of eider (*Somateria mollissima*) and long-tailed duck (*Clangula hyemalis*). *Zool. Revy* **35**: 89–92. (In Swedish with English summary.)

Pehrsson, O. (1974). Nutrition of small ducklings regulating breeding area and reproductive output in the long-tailed duck, *Clangula hyemalis*. *Int. Congr. Game Biol.* **11**: 259–64.

Pehrsson, O. (1975). Regional, seasonal, and annual fluctuations of the goldeneye, *Bucephala clangula* (L.) on the Swedish west coast. *Viltrevy. Swedish Wildlife* **9**(6): 239–302.

Pehrsson, O. (1976a). Estuarine production of waterfowl food on the Swedish west coast. In *Fresh water on the sea*, pp. 221–6. The Association of Norwegian Oceanographers, Oslo.

Pehrsson, O. (1976b). Food and feeding grounds of the Goldeneye *Bucephala clangula* (L.) on the Swedish west coast. *Ornis Scand.* **7**: 91–112.

Pehrsson, O. (1976c). The importance of food in the regulation of some diving duck populations. PhD thesis, University of Göteborg, Göteborg, Sweden.

Pehrsson, O. (1978). A ten-year fluctuation pattern of the common eider (*Somateria mollissima*) on the Swedish west coast as a result of food availability. In *Proceedings from the symposium on sea ducks*, pp. 91–8. SNV (National Swedish Environment Protection Board) PM 1009.

Pehrsson, O. (1979). *Grunda kustområdens betydelse för förekomst av sjöfågel och säl – sammanställning och bearbetning av ett inventeringsmaterial för norra Halland och Bohuslän.* Natur och kulturvårdsprogram för Göteborg. 6 pp. (In Swedish.)

PART 2

Bird populations: social behaviour and the use of feeding areas

Introduction

J. D. GOSS-CUSTARD.

The response of birds to spatial and seasonal variations in the abundance of their food may be considerably modified by their reactions to other birds, of the same or of different species. In terms of their effect on the individual, these responses may have either advantageous or disadvantageous consequences. This is because the environment places contradictory demands on organisms which have to be resolved by individuals through compromise. For instance, many shorebirds may benefit from foraging in a flock, perhaps because this reduces the risk from predators. But in some species, crowding together may cause birds to interfere with each other's feeding so that the rate of food intake decreases and the risk of starvation increases. The feeding dispersion which the birds then adopt in response to these opposing demands may represent a compromise which varies according to how prone the birds' feeding method is to interference (Goss-Custard 1970): the more susceptible species may feed in loose flocks, even though the risk from predation is thereby increased, whereas others may be able to forage in tight flocks without a serious reduction in intake rate. Viewing adaptations in terms of their respective advantages and disadvantages, or costs and benefits, has become widespread in studies of the adaptive significance of behaviour and has proved to be an effective way in which to view adaptations.

This approach has been adopted by several authors in this part of the book. Ydenberg & Prins review new ideas and studies on the benefits to be derived from roosting communally at high water. These include a measure of protection from inclement weather and, perhaps, from predators though, as they point out, the evidence for this is very weak. However, individuals in communal roosts may benefit by not having to maintain such a high level of vigilance as would birds roosting singly, and this may have a number of benefits. But the most striking association they detect is that between communal roosting and communal foraging on patchy and unpredictable food supplies on which birds may feed in high densities without the risk of serious interference. Ydenberg & Prins argue that the information-centre hypothesis proposed by Ward & Zahavi (1973) may account for communal roosting in waders, and cite some experimental evidence in favour of this. They note that shorebirds provide good subjects for further work on this issue.

On a number of occasions, Ydenberg & Prins point out that the benefits of roosting communally may be offset by certain costs, at least in some individuals. Thus subordinate birds may be pushed to the edge of the roost in windy weather and so bear the brunt of the gale. However, despite this extra cost, these individuals all retain the presumably overriding benefits of remaining in the flock. But this example does illustrate how the balance between costs and benefits may vary between individuals, a point that is thought to be increasingly important in understanding the function of various patterns of behaviour.

This theme is taken further by Townshend, Dugan & Pienkowski in their chapter on foraging Grey Plovers. As in several shorebird species, some individuals forage in territories while others do not. But interestingly, they discover that, amongst the territorial individuals, on the Tees estuary, some occupy their territories for several months while others occupy them for only a matter of days or weeks. Accordingly, separate comparisons have to be made between the costs and benefits associated with each of three options. Birds which do not occupy territories at all may benefit from being able to range more widely over the estuary and so be able to exploit local patches of food abundance particularly, it seems, at night. Those occupying short-term territories share this benefit but may also gain from reduced rates of interference at critical times in the winter when energy demands are most difficult to meet. The holders of long-term territories have this latter benefit at all times, but also may gain by having a longer lasting food supply, achieved through slower rates of depletion and by the conservation of the best patches of prey until they are desperately needed. They may also benefit by having available, in their territories, deep creeks within which they can shelter in winter gales. To set against these advantages, the birds adopting the three options may experience different costs. Those in territories may be more prone to predation, and those in long-term territories may be least likely to discover new and profitable sources of food as they occur. The non-territorial birds, in contrast, may be less prone to predation but more likely to suffer reduced intake from prey depletion through the winter.

This is an intriguing discussion of a complex situation, and it serves to illustrate how many considerations may need to be taken into account if we are properly to understand the behavioural adaptations shown by shorebirds. Townshend, Dugan & Pienkowski discuss the significance of the diversity of individual responses to the food supply, and to other birds, on the Tees. They note the possibility that the three options may be, in the long term, equally successful for different birds, but not necessarily so for any one individual. They conclude that some birds may be prevented from obtaining long-term territories because of the presence of other, more dominant, individuals, but that some individuals may choose to feed non-territorially. Thus competition between birds need not necessarily be of paramount importance in determining use of feeding space on the Tees.

The final chapters in this section present evidence that competition on the feeding grounds does occur in some species on the migration and overwintering areas. This is an area of research which owes much to the pioneering studies by Leo Zwarts on the so-called 'buffer-effect'. Van der Have, Nieboer & Boere show that juvenile Dunlin, arriving in the Dutch Wadden Sea, are represented disproportionately in the inshore feeding areas. These appear to be the least preferred because bird density there is relatively low. Most adults occur in those regions of the Dutch Wadden Sea which are occupied first and where bird densities are typically high. Goss-Custard & Durell report almost identical results from a study of Oystercatchers in a small estuary in southwest England. In their study, however, they were able to show that the young birds were sub-dominant to adults and vacated preferred areas as the adults arrived, but returned to them in the spring when the adults left for the breeding areas.

Although both these chapters present data which suggest that the low-status, young birds may experience a disproportionate mortality in winter, in neither was the link established strongly. Swennen's study on Oystercatchers in the Dutch Wadden Sea provides further evidence for both competition and differential rates of mortality occurring in different sections of a population. Oystercatchers using small roosts adjacent to narrow, truncated feeding areas (which were rendered unavailable to the birds in storms) suffered a much higher mortality than those birds roosting next to large feeding areas (which were usually available, however bad the conditions). All the roosts, and their associated feeding grounds, were near to each other, so it is highly unlikely that birds from the small roosts could not have found the better feeding areas. It seems much more likely that competition occurred on the best feeding areas, and that this kept out those birds found at the small roosts. This implies that these birds were of low status, and this is indicated by Swennen's finding that a high proportion of these birds had severe abnormalities of one kind or another. Similarly, Goss-Custard & Durell showed that low-status Oystercatchers were more likely to feed in the least preferred mussel beds of the Exe.

There does seem, therefore, to be an increasing amount of evidence to suggest that competition occurs on the migration and wintering grounds. The key question, though, is whether the sub-dominant birds are thereby more likely to die as a result. Both van der Have, Nieboer & Boere and Goss-Custard & Durell make the point that a greater risk of mortality in the sub-dominants may reflect a lack of experience in feeding, rather than a failure to compete with other birds. Since young birds are generally sub-dominant to adults, inexperience in feeding and low status tend to go together. It is therefore important for us to disentangle the extent to which competition and inexperience contribute to mortality, so that we may properly appreciate the cost of failure which the losers bear. As Goss-Custard & Durell mention, this cost could be quite trivial.

The importance of this issue to applied research is stressed in the final chapter by Goss-Custard & Durell. If competition occurs, and it contributes to any difficulties some birds have in collecting enough food, mortality will be density-dependent. If feeding areas on estuaries are lost through reclamation and industrial development, competition in the reduced feeding areas will increase, and so will mortality. Even if most of the annual mortality, as well as all the production, takes place in the summer, an increase in winter mortality in the young birds may substantially reduce the size of the population as a whole. Studying the extent to which competition for food occurs is therefore not only of significance in drawing up the balance sheet in a cost–benefit analysis of behavioural function: it may be of some considerable practical significance as well.

References

Goss-Custard, J. D. (1970). Feeding dispersion in some over-wintering wading birds. In *Social behaviour in birds and mammals*, ed. J. H. Crook, pp. 3–35. Academic Press, London.

Ward, P. & Zahavi, A. (1973). The importance of certain assemblages of birds as 'information-centres' for food finding. *Ibis* **115**: 517–34.

7

Why do birds roost communally?

R. C. YDENBERG & H. H. Th. PRINS

Introduction

In this article we explore the adaptive significance of communal roosting. We have adopted the approach of drawing on a wide variety of studies in order to examine in some detail a few of the ideas which may help us to understand why so many waders and wildfowl roost in large groups.

Throughout the writing of this article, we have kept in mind the question 'Why should an individual bird decide to roost with other birds?' Although this only rephrases the question of why birds roost communally, we have three reasons to believe that this is a particularly appropriate way to ask it.

Two of these reasons relate to function. First, the evidence is that natural selection acts overwhelmingly at the level of the individual, and so any way of thinking about an evolutionary problem in this way can be a reliable guide to intuition. Second, since we ask 'why', we stress that we are asking a question about function, rather than mechanism. These two – while closely related – should not be confused as alternative explanations for behaviour. The literature on roosting frequently uses the idea of tradition to explain roosting habits such as the choice of roost site. Tradition is a mechanism whereby birds acquire certain behaviours, namely by social experience; it does not explain why they acquire these habits in this way.

The third reason is that we have assumed that the decisions to join roosts are made by individuals in a rational and economic (though not necessarily conscious) manner. A great deal of progress has been made in the last decade in illuminating the behaviour of animals using this economic, or cost-benefit approach (Krebs & Davies 1978). The recent excellent studies of Myers, Connors & Pitelka (1981) and Ens & Zwarts

(1980) have convincingly shown that economic models can help to explain why many waders are territorial on their wintering areas.

In formulating such an economic theory, we must turn to two traditional means of investigation in order to discover the selective forces involved in the maintenance of communal roosting. In the following sections we first analyse the results of several comparative studies and then review some studies of particular species, concentrating on quantitative and experimental studies.

Comparative studies

The most salient feature of communal roosting is its association with social feeding. Whereas only about a quarter of all birds that nest in colonies feed socially (Lack 1968), very few communal roosters do not (Ward & Zahavi 1973). Some species such as tits (Paridae) and sandgrouse (Pteroclidae) feed in flocks and roost solitarily, but this is rare. Other solitary feeders (dippers (Cinclidae); treecreepers (Certhidae)) may roost together only in very cold weather. It seems that the factors which make it advantageous for individuals to join communal roosts tend to occur together with factors that also favour the habit of feeding in flocks.

But what is the nature of this link between social feeding and communal roosting? This is just the sort of question that a comparative study can help to answer. By surveying groups of species we may identify characteristics shared by communal roosters.

Two studies of passerine groups are particularly helpful here. Newton (1972) investigated the winter roosting habits and diets of finches in the British Isles. He found that species which exploit seeds from the ground are birds of the open country, wandering great distances daily and gathering into large, permanent communal roosts. In species whose food occurs in scattered patches of limited size, such as cones on trees or seeds on plants, roosts are much smaller, temporary, and all the birds in a single roost usually feed together. The only finches to roost and feed alone are those Chaffinches (*Fringilla coelebs*) which occupy territories. Non-territorial Chaffinches, such as those that immigrate to Britain during the winter, feed on the ground and wander over large areas and, as expected, feed and roost in large groups. Coombs' (1978) study of the European corvids reveals similar trends (Table 7.1). Rooks and Jackdaws are seed and grain specialists during winter and gather into large roosts. In contrast, species that remain on their territories during the winter sleep alone or in small family groups and rely on hoarded food. Species that exploit ephemeral patches of limited size, such as carrion,

Table 7.1. *Roosting and winter food habits of the European Corvidae*

Species	Roost size	Winter diet
Perisoreus infaustus (Siberian Jay)	Solitary/pairs, small groups	Stored nuts and berries on year-round territories
Garrulus glandarius (Jay)	Solitary/pairs	Stored nuts on year-round territories
Nucifraga caryocatactes (Nutcracker)	Solitary/pairs	Stored nuts on territories
Pyrrhocorax pyrrhocorax (Chough)	Solitary/pairs	Insects, especially ants; remains on territories
Pica pica (Magpie)	Usually 20, occasionally more	Insects, grubs, seeds
Pyrrhocorax graculus (Alpine Chough)	25–30	Berries
Corvus corax (Raven)	Up to 150	Carrion, opportunistic predation, miscellaneous vegetative matter
Corvus carone (Carrion Crow)	Up to 200	Carrion, opportunistic predation, grain and seeds
Corvus monedula (Jackdaw)	100s–1000s	Grain and seeds
Corvus frugilegus (Rook)	100s–1000s	Grain

From Coombs (1978).

roost in groups of intermediate size. The clear indication from these two studies is that the dispersion of food is an important factor in determining roosting behaviour.

However, this conclusion should be treated with caution. When we observe that some trait is correlated with a variable, it is tempting to conclude that there is a causal connection between the two. However, ecological variables are often associated with one another. For example, among the terns (Laridae) of both the Atlantic and Pacific Oceans, body size, foraging range, flock size, food size selection and colony size are all closely correlated (cf. Ashmole 1968; Erwin 1979). On the North American Pacific coast, the body size, foraging range, diving capability, chick growth rate and the colony size of alcids (Alcidae) are all strongly interrelated (Cody 1973). How can we separate the effects of all these variables? Mace (1979) tackled the problem statistically. She surveyed the relative brain size of genera of rodents and found that it correlated well with habitat, diet, stratification (i.e., arboreal, terrestrial, fossorial) and the timing of periods of activity. When the effects of other variables were held constant statistically, only the relationship with diet remained.

A second problem is choosing the taxonomic level to provide the unit of observation (Clutton-Brock & Harvey 1979). In the case of the corvids, for example, the result might be biased because four of the communal roosters belong to the genus *Corvus*. The outcome might be different if genera were used as the basis of comparison.

In our comparative survey of the Anatidae, we were unable to gather enough information for statistical treatment. Instead, we compared trends at two taxonomic levels because, if both show the same pattern, there is less need to be concerned about the problem of independence (Krebs 1978).

Most of the data come from two comprehensive works on the Anatidae (Todd 1979; Johnsgard 1978). The diets and roosting habits of 40 of the 41 genera were sorted into several broad categories. Usually this was not too difficult, although in some cases the judgement was based on very little information. Birds that roosted in pairs or family groups were not classed as communal roosters. The analysis was repeated for the genus *Anas*. This gave two comparative reviews of nearly equivalent size at two taxonomic levels.

The results are summarized in Fig. 7.1. The same broad trends are apparent in both taxonomic groups: those species exploiting vegetable matter tend to congregate in roosts, while those consuming animals are far more likely to roost alone.

We suggest that the reasons for this association are much the same as those proposed for the Corvidae and Fringillidae, though they are perhaps more complex. The seeds eaten by waterfowl are usually small and must be sieved from the substrate or from the water. They usually occur in large but scattered and inconspicuous patches (see van Eerden, Chapter 5 of this volume). Although at first sight the forage of grazing waterfowl seems to be distributed much more evenly than seeds, much of it is unsuitable because the birds must consume only high quality material in order to maintain a positive nitrogen balance. Even during the winter they may rely on new growth and this occurs unevenly according to local conditions. Indeed they may congregate in order to concentrate their grazing, thus keeping plants growing when they might otherwise slow down or stop (Prins, Ydenberg & Drent 1980; Ydenberg & Prins 1981). In either case, the food occurs in large and scattered patches.

Unlike these herbivores, carnivorous waterfowl eat food that is in many cases able to move. This has two important consequences. First, much of the standing crop may be unavailable to foraging birds: the data of both Zwarts & Wanink (Chapter 4 of this volume) on *Mya* eaten by Curlew and Goss-Custard (1970) on *Corophium* eaten by Redshank suggest that over 99% of the prey may simply be unobtainable at any one time. Animal food may therefore be much more thinly distributed than plant food. Second, carnivores may interfere with each others' foraging success as prey animals may hide when they detect predators (Goss-Custard 1970). In these cases, the reduction in intake rate from interference may make communal feeding too costly.

That the trends in Fig. 7.1 are so similar inspires confidence, but there remains the deeper problem of confounding variables which can apply at both taxonomic levels. We could not deal with this problem statistically, but can remove one factor that has confounded other studies. Among the dabbling ducks (*Anas* spp.), body size is not associated with any particular type of roosting dispersion. Small, medium and large ducks are all equally likely to roost communally: 58% of the 12 species with a body weight of 500 g or less roosted communally compared with 70% of 10 species weighing 500–750 g and 62% of 13 species weighing over 750 g. We can summarize by saying that the comparative survey

Fig. 7.1. The winter diets and communal roosting habits of (*a*) 40 genera in the Anatidae (no data available for *Callonetta*), and (*b*) 35 species in the genus *Anas*.

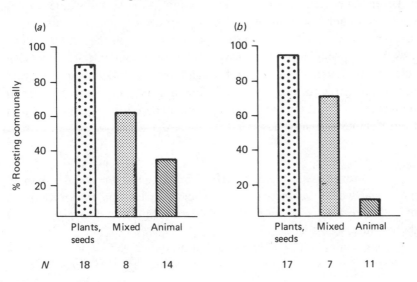

supports the notion that communal roosting is strongly influenced by diet, but we could not confirm this conclusion in a statistical way.

Observational and experimental studies

This section reviews studies of specific selection pressures which may increase or decrease the fitness of individuals joining roosts. The list is not exhaustive; instead we focus on a few ideas which may be particularly appropriate for waterfowl.

Roost microclimate

Communal roosts may be warm places to sleep, owing either to the physical characteristics of the roost site itself (Siegfried 1971; Kelty & Lustick 1977) or to the presence of many birds, warming the roost above ambient temperature. Whitlock (1979) has shown in experiments with models that the windspeed is lower in the centre of Redshank roosts than at the edge. Thus individuals may profit by joining communal roosts because they have to expend less energy to stay warm.

But how big is the energy saving? Both Yom-Tov, Imber & Otterman (1977) studying *Sturnus vulgaris* in Israel and Gyllin, Källander & Sylven (1977) working on *Corvus monedula* in Sweden have measured the microclimatic advantage gained by individual birds in large communal roosts and concluded that although it is substantial, it is never sufficient to offset the costs of the long distances flown by some birds to join these roosts. The microclimate advantage is not in itself enough to account for the formation of communal roosts.

This is supported by another observation. As conditions deteriorate, adult Redshank force juveniles to the windward side of roosts, thus usurping for themselves the more protected spots in the centre (Whitlock 1979). Swingland (1977) saw similar behaviour in rook roosts on cold nights. Juveniles may therefore gain no energetic advantage at all at the most stressful times. Indeed, they may actually 'pay' in order to remain in the roost, suggesting that some benefit other than protection from weather is involved.

Predation

Birds in groups may be safer from attacks by predators than an individual would be on its own. Several reviews on this have been published recently (Bertram 1978; Pulliam & Millikan 1983), so here we discuss just two aspects.

Page & Whitacre (1975) showed that predation can be heavy in small waders: 7.5 % or more of four species wintering in Bolinas Lagoon,

California, were killed by raptors during one winter. Four other species had *known* losses, based on carcass recoveries, of at least 5% and the true losses were probably substantially higher.

For these birds, the critical element in escaping successfully was the timing of taking flight. Of 234 attacks on birds on the ground, 15.6% were successful, while none of the 92 attacks on flying birds resulted in a kill. For this reason, it is often supposed that individuals in large groups of birds are at an advantage – they spot approaching predators sooner and so take flight more quickly. This has been summed up in the simple model devised by Pulliam (1973), and the experiments of Powell (1974), Siegfried & Underhill (1975) and Kenward (1978) are frequently cited in support.

However, birds may actually have seen the approach of a predator but failed to react immediately because there is a cost to responding too soon. Page & Whitacre (1975) suggested that small waders need only get into the air to evade Merlins (*Falco columbarius*) and Kenward (1978) states that healthy Woodpigeons (*Columba palumbus*) can easily outfly

Fig. 7.2. The flight distance of Barred Ground Doves (*Geopilia striata*) from an approaching predator as a function of flock size. The relationship is best described by a quadratic equation ($y = 10.76 - 0.63x + 0.11x^2$). Sample sizes for flock sizes up to four are all greater than 30, but for flocks greater in size than five the sample sizes are all four. The vertical lines represent ± 1 s.D. From Grieg-Smith (1981), reproduced here by permission.

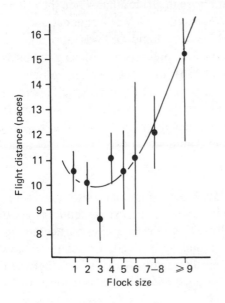

their Goshawk (*Accipiter gentilis*) predators. For these birds, an early response is unnecessary and occurs at the cost of losing feeding time. As a predator approaches, a foraging bird should weigh up the benefit of remaining to feed against the risk of too late a departure. If the rate of feeding varies with flock size (due to interference, for example), the reaction distance to approaching predators may also change with flock size – not because birds in smaller groups have not sighted the predator, but because the benefit of remaining to feed when interference is low makes the risk of a delayed take-off worthwhile.

There is some evidence in support of this interpretation. Grieg-Smith (1981) has shown that the reaction distance of Barred Ground Doves (*Geopilia striata*) to an approaching predator (himself) was smallest in groups of three birds (Fig. 7.2), and he pointed out that the data of Siegfried & Underhill on the related Laughing Dove (*Streptopelia senegalensis*) show a similar inflection. He interpreted this to mean that the benefit of feeding was highest in groups of three.

Even though birds in large groups may not actually benefit from improved predator detection, they gain from being able to devote more time to feeding than they safely could in smaller groups (Pulliam 1973; Caraco, Martindale & Pulliam 1980). The significance of this for communal roosting is that birds roosting in large groups may be able to sleep more. Birds normally sleep in bouts of rather short duration, interrupted by 'peeks', in which they open their eyes and scan the surroundings. Lendrem (1983) has found that Barbary Doves (*Streptopelia risoria*) sleep more when they are in groups than when they sleep alone (Fig. 7.3). Doves in groups 'peeked' at a lower frequency and as a result, slept in longer bouts. Moreover, Lendrem was also able to manipulate the distribution of inter-peek intervals by exposing the birds to a predator (a polecat-ferret on a leash): the doves peeked more often after having seen the polecat. We assume that there is some benefit to sleeping longer or more deeply, but are unable to identify this as yet. It is feasible that the magnitude of this benefit is great enough to explain why birds come from so far to join communal roosts.

Information centres

Many writers have speculated that communal roosts are centres which enable individuals to learn in some way the whereabouts of good feeding sites. The 'information-centre' hypothesis applies particularly to species that exploit unpredictable and short-lived, or scattered and hard-to-find, food resources and so fits neatly with the link between food supplies of this kind and the occurrence of communal roosting. A number of investigations have been made recently, both in the field and laboratory.

(a) Field studies. Field investigations at colonies and roosts have concentrated on finding out whether birds actually follow one another to good feeding sites. Herons (*Ardea* spp.) in coastal colonies often show a marked synchrony in the timing of foraging flights from the colony (Krebs 1974). This is consistent with the information-centre hypothesis, but synchrony may be imposed by external factors such as the tide or

Fig. 7.3. The distribution of 'inter-peek' intervals of Barbary Doves under various experimental conditions. FS = flock size; +P = predator seen just prior to experiment; −P = no predator. The thick lines are the observed and the thin lines the expected distributions; the expected distribution was generated from the exponential. Very short bouts are overrepresented in small flocks and when the predator had been seen, but approximate to a random distribution in larger flocks. The overabundance of short bouts is maintained in larger flocks when the predator was present. From Lendrem (1983).

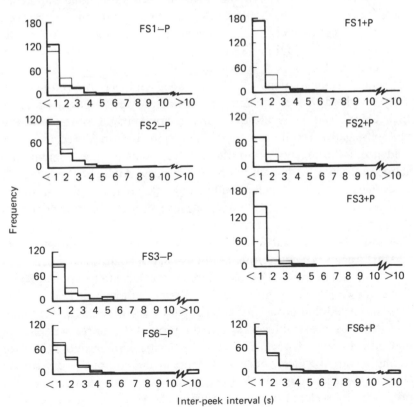

Inter-peek interval (s)

cycles in the availability of food (cf. Evans 1981). However, Krebs (1974) found that the synchrony within subdivisions of a colony of the Great Blue Heron (*Ardea herodias*) was greater than those between subdivisions, suggesting that external factors were not entirely responsible. Furthermore, although the direction taken by departing adults varied widely, the members of subdivisions within the colony were more likely to take the same route than would be expected by chance. These observations accord well with an information-centre interpretation, although the evidence is, of course, circumstantial.

In two studies bait was placed at stations in the field and the recruitment of birds to them recorded. Loman & Tamm (1980) put carcasses of pigs and chickens out during the winter in Sweden, counted the number of crows and ravens that discovered them during the course of the day, and then counted the birds that returned the following morning. They reasoned that if birds learned about these desirable food sources from conspecifics at roosting sites, greater numbers of birds than observed on the previous day would return to the carcasses. Although their results are inconclusive because no birds were ringed and they could not tell how many individuals found the carcass on the first day, they are suggestive (Fig. 7.4). Though crows did not conform to the expectation, ravens did.

Andersson, Gotmark & Wicklund (1981) baited rafts with fish around an island-breeding colony of Black-headed Gulls (*Larus ridibundus*). They observed the return trips to the raft of over 50 parents feeding their chicks, and never observed any instances of following. Unfortunately, negative results such as these cannot be considered as strong evidence: following may occur only rarely, or only in times of great food shortage, or may be difficult to observe directly. Although Andersson *et al.* believe that Black-headed Gull colonies do *not* function as 'information-centres', their results, though suggestive, are as circumstantial as the positive evidence discussed above.

A more powerful test would be to observe the behaviour of unsuccessful foragers. Of course this would be very difficult, since the identity *and* the recent nutritional status of birds would have to be known. So far as we know, no one has attempted to do this, although several workers have suggested that unsuccessful birds hang back at roosts and colonies to watch for successful feeders. The duration of the morning departure of Cattle Egrets (*Ardeola ibis*) in South Africa (Siegfried 1971), for example, varied widely, but was longest at the end of the dry season (Fig. 7.5) when food was in shortest supply. Siegfried speculates that the prolonged roost departure at this time of the year is due to more birds

Fig. 7.4. The numbers of crows and ravens visiting carrion put out in winter in Sweden. Counts 1–4 represent successive tallies of birds at the bait at 2-hour intervals after the carrion was discovered. After roosting overnight, more ravens return to the bait than the maximum numbers counted on the previous evening, but the number of crows returning is equal to the number previously observed. Plotted are the 1977 data given in Loman & Tamm (1980), but the trends for their 1978 data are similar.

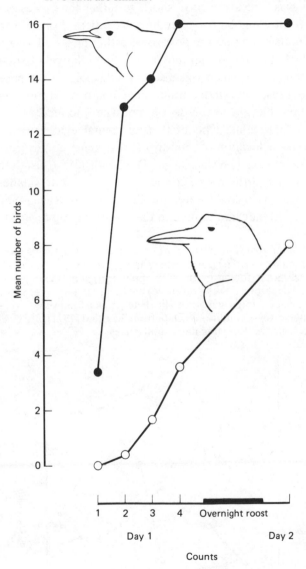

watching for clues as to the whereabouts of foraging sites better than those they used the previous day.

In a study on wintering Barnacle Geese (*Branta leucopsis*), we found that post-roost gatherings lasted longer on mornings following days on which geese had foraged until very late (Fig. 7.6). An examination of the factors associated with late departures from the foraging grounds suggested that this was a result of food shortage rather than a case of resting later the day after a good day's feeding (Ydenberg, Prins & van Dijk 1983). We hypothesized that the post-roost gathering lasted longer when more individuals had to consider moving to other relatively distant feeding areas, the quality of which was also unpredictable. After poor foraging days, the geese invested considerable amounts of time in gathering information relevant to the decision of where to feed.

(b) Laboratory experiments. The first experimental evidence that birds can learn about the location of food from roost-mates comes from a laboratory experiment on *Quelea quelea*. De Groot (1980) used an aviary consisting of a central roosting room joined to four smaller 'resource' rooms. The birds could enter the resource rooms via small tunnels in the walls, and they had to fly into these rooms to discover the

Fig. 7.5. Communal roost departures of the Cattle Egret (*Ardeola ibis*) during the annual cycle. The duration is longest at the end of the dry season (March) when food is in shortest supply. Intriguingly enough, the size of the departing flocks (top curve) also peaks at this time, as would be expected if birds were more likely to adopt a following strategy under adverse conditions. Data from Siegfried (1971). Rainfall: open circles; Roosting data: solid circles.

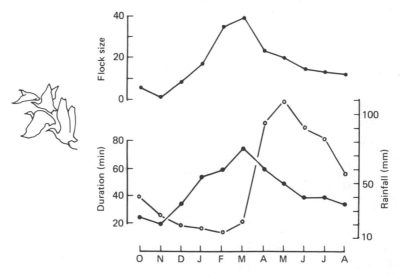

resource they contained. De Groot trained one group of six birds to find food in one of the resource rooms, and trained a second group to find water in another. The groups were allowed to roost together, and were deprived of food. After several hours, they were given access to the resource rooms and the number of naive birds (i.e. those ignorant of the location of food, in this case) entering the correct resource room was counted. The birds were then deprived of water for several hours and the experiment repeated. The results from five such trials were clear-cut: the naive birds could always find the correct resource rooms. In control experiments, De Groot proved that the naive birds had gained their information from their knowledgeable roost-mates, since they were incapable of finding the correct room on their own, even when other birds were feeding noisily there. In a second experiment, naive birds showed that they were capable of choosing the more profitable of

Fig. 7.6. The duration of the post-roost gathering of Barnacle Geese (*Branta leucopsis*) shows a strong positive relation with the lateness of the departure from the foraging grounds on the preceding evening. Late departures from the feeding area seem to be due to poor foraging conditions. From Ydenberg *et al*. (1983). $Y = 2.85x - 183$; $n = 22$; $F_{slope} = 25.2$, $P < 0.001$.

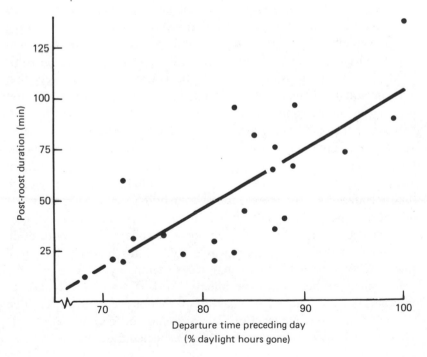

two similar resources (pure seed versus seed mixed with sand) after roosting with knowledgeable companions.

How did the naive birds gain their information? Although the knowledgeable birds tended to enter the resource rooms first (thereby suggesting that the naive birds may have followed them), this was significant in only 3 of 15 experiments. In one trial, four naive birds entered the correct chamber before any of the informed individuals! Although De Groot feels that the knowledgeable birds indicated which tunnels led to food or water by repeatedly flying to them before access was allowed, he collected no data on this.

Dispersed roosting

So far we have considered only factors that might promote the tendency of birds to roost together. There are circumstances, however, in which individuals may not join roosts; even highly gregarious species like Starlings occasionally roost alone or in pairs during the non-breeding season (Clergeau 1981). Condors (*Gymnogyps californianus*) often do not return to communal roosts when they have found a carcass, but roost nearby until it is consumed (Koford 1953): perhaps returning to the communal roost would result in unwelcome sharing of the carrion. Other Condors may be able to tell which individuals have been feeding well (by their laboured flight for example), and so identify which birds to follow. But not using the communal roost may be risky in small species subject to heavy predation or chilling. In these cases, we would expect selection to favour 'concealment mechanisms' within the roost itself, though this possibility has not been tested. It is unlikely that animals would obligingly share information about the location of food if it meant increased competition.

Another factor that may affect group size is the variance in foraging success. In a simulation investigating the effects of flock size on the foraging success of Black-capped Chickadees (*Parus atricapallus*) in winter, Thompson, Vertinsky & Krebs (1974) found that the flock size that maximized average food intake (four birds) was smaller than that minimizing the variance (16 birds). In large flocks the birds had very little chance of starving since large flocks discovered food more often than small ones. However, there was also very little chance of doing well since any food discovered was shared between many birds.

Recent theoretical and experimental work (Caraco, Martindale & Whittam 1980; Stephens 1981) has shown that birds should be sensitive not only to the average level of the distribution of food, but also to its variance. The models predict that when expected food intake (over 24 hours) exceeds the requirement, birds should 'play safe' by choosing safe foraging tactics. When the projected energy budget is negative so that starvation becomes a threat, birds must gamble on more lucrative tactics, but at the risk of doing very poorly. In an experiment with Dark-eyed Juncos, this is exactly what Caraco *et al.* (1980) found.

These results suggest that solitariness may be a 'risk-prone' strategy to which birds are forced to resort in times of stress. It is interesting that solitary Woodpigeons are in poorer condition than flock birds (Kenward 1978). Could it be that these birds are behaving in a risk-prone manner to increase their chances of surviving? Though so far applied only to foraging flocks, these results have obvious implications for the decisions of birds to join communal roosts.

Conclusion

This article has concentrated on questions of why – functional questions. It has explored the evolution of communal roosting by reviewing both comparative and single species studies. The comparative surveys show a correlation between communal roosting, diet and food dispersion in several groups of birds, but there are some reasons to be cautious about drawing firm conclusions from this. The experimental evidence shows that microclimatic considerations and the information-centre function of communal roosts are probably important, but the field evidence for the information-centre function of roosts is circumstantial. The evidence for the anti-predation hypothesis remains equivocal.

Communal roosting is an interesting evolutionary problem worthy of further study, and there are several areas in which profitable research may be undertaken. A comparative review of the waders might reveal trends different to those shown by the wildfowl. There is still a need for more information on roosting itself. Several experimental approaches suggest themselves, and waders and wildfowl may be good subjects for such work. There are many ringed birds in some populations and their haunts are well known and relatively accessible. Also, a great deal of effort, expense and ingenuity has already been applied in devising techniques to observe and follow these birds in the field, and to measure the things that they do.

Acknowledgements

We thank Peter Grieg-Smith and the editors of *Ibis*, the journal of the British Ornithologists Union for permission to use Fig. 7.2, which originally appeared in *Ibis* vol. 123. We also thank Alain Tamisier, Bart

Ebbinge, Bruno Ens and Vanessa Power for the discussion of some points and for supplying unpublished data. Bill Sutherland, Dennis Lendrem, John Horsfall and John Krebs read earlier versions of this manuscript and helped to improve it. Dennis Lendrem also drew Fig. 7.3. We are especially grateful to Maggie Birkhead for drawing the vignettes.

References

Andersson, M., Gotmark, F. & Wicklund, W. (1981). The information-centre function of black-headed gull (*Larus ridibundus*) colonies. *Behav. Ecol. Sociobiol.* **9:** 199–202.

Ashmole, N. P. (1968). Body size, prey size and ecological segregation in five sympatric tropical terns (Aves: Laridae). *Syst. Zool.* **17:** 292–304.

Bertram, B. C. R. (1978). Living in groups: predators and prey. In *Behavioural ecology*, ed. J. R. Krebs & N. B. Davies, pp. 64–96. Blackwell, Oxford.

Caraco, T., Martindale, S. & Pulliam, H. R. (1980). Avian flocking in the presence of a predator. *Nature* **285:** 400–1.

Caraco, T., Martindale, S. & Whittam, T. (1980). An empirical demonstration of risk-sensitive foraging preferences. *Anim. Behav.* **28:** 820–30.

Clergeau, P. (1981). 'Non-roosting' starlings in Brittany. *Ibis* **123:** 527–8.

Clutton-Brock, T. H. & Harvey, P. H. (1979). Comparison and adaptation. *Proc. roy. Soc., Lond.* (*B*) **205:** 547–65.

Cody, M. L. (1973). Co-existence, co-evolution and convergent evolution in seabird communities. *Ecology* **54:** 31–44.

Coombs, F. (1978). *The crows: a study of the corvids in Europe.* B. T. Batsford Ltd, London.

De Groot, P. (1980). Information transfer in a socially roosting weaver bird (*Quelea quelea:* Ploceinae): an experimental study. *Anim. Behav.* **28:** 1249–54.

Ens, B & Zwarts, L. (1980). Territoriaal gedrag bij wulpen buiten het broedgebied. *Watervogels* **5:** 155–69.

Erwin, R. M. (1979). Coloniality in terns: the role of social feeding. *Condor* **80:** 211–15.

Evans, P. G. H. (1981). Ecology and behaviour of the Little Auk *Alle alle* in west Greenland. *Ibis* **123:** 1–18.

Goss-Custard, J. D. (1970). Feeding dispersion in some over-wintering wading birds. In *Social behaviour in birds and mammals*, ed. J. H. Crook, pp. 3–35. Academic Press, London.

Grieg-Smith, P. W. (1981). Responses to disturbance in relation to flock size in foraging groups of barred ground doves *Geopilia striata. Ibis* **123:** 103–6.

Gyllin, R., Källander, H. & Sylven, M. (1977). The microclimate explanation of town centre roosts of jackdaws *Corvus monedula. Ibis* **119:** 358–61.

Johnsgard, P. A. (1978). *Ducks, geese and swans of the world.* University of Nebraska Press, Lincoln.

Kelty, M. P. & Lustick S. I. (1977). Energetics of the starling *Sturnus vulgaris* in a pinewood. *Ecology* **58:** 1181–5.

Kenward, R. E. (1978). Hawks and doves: attack success and selection in goshawk flights at woodpigeons. *J. anim. Ecol.* **47:** 449–60.

Koford, C. B. (1953). The California Condor. *National Audubon Society Research Report* **4.**

Krebs, J. R. (1974). Colonial nesting and social feeding as strategies for exploiting food resources in the great blue heron (*Ardea herodias*). *Behaviour* **51:** 99–134.

Krebs, J. R. (1978). Colonial nesting in birds, with special reference to the Ciconiiformes. *National Audubon Society Research Report* **7:** 299–314.

Krebs, J. R. & Davies, N. B. (1978). *Behavioural ecology.* Blackwell, Oxford.

Lack, D. (1968). *Ecological adaptations for breeding in birds.* Methuen and Co., London.

Lendrem, D. W. (1983). Vigilance in birds. D. Phil. thesis. University of Oxford, England.

Loman, J. & Tamm, S. (1980). Do roosts serve as 'information-centres' for crows and ravens? *Am. Nat.* **115:** 285–9.

Mace, G. M. (1979). The evolutionary ecology of small mammals. D.Phil. thesis, University of Sussex, Brighton, Sussex, England.

Myers, J. P., Connors, P. G. & Pitelka, F. A. (1981). Optimal territory size and the sanderling: compromises in a variable environment. In *Foraging behaviour: ecological, ethological and psychological approaches*, ed. A. Kamil & T. Sargent, pp. 135–58. Plenum Press, New York.

Newton, I.'1972). *Finches.* William Collins & Sons, London.

Page, G. & Whitacre, D. F. (1975). Captor predation on wintering shorebirds. *Condor* **77:** 73–83.

Powell, G. V. N. (1974). Experimental analysis of the social value of flocking by starlings (*Sturnus vulgaris*) in relation to predation and foraging. *Anim. Behav.* **22:** 501–5.

Prins, H. H. Th., Ydenberg, R. C. & Drent, R. H. (1980). The interaction of Brent Geese *Branta bernicla* and sea plantain *Plantago maritima* during spring staging: field observations and experiments. *Acta Bmo. Nederl.* **29:** 585–96.

Pulliam, H. R. (1973). On the advantages of flocking. *J. Theor. Biol.* **38:** 419–22.

Pulliam, H. R. & Millikan, G. C. (1983). Social organisation in the non-breeding season. *Avian Biol.* **6:** 169–97.

Siegfried, W. R. (1971). Communal roosting of the cattle egret. *Trans. roy. Soc. S. Afr.* **39:** 419–43.

Siegfried, W. R. & Underhill, L. G. (1975). Flocking as an anti-predator strategy in doves. *Anim. Behav.* **23:** 504–8.

Stephens, D. W. (1981). The logic of risk-sensitive foraging preferences. *Anim. Behav.* **29:** 628–9.

Swingland, I. R. (1977). The social and spatial organisation of winter communal roosting in rooks (*Corvus frugilegus*). *J. Zool., Lond.* **182:** 509–28.

Thompson, W. A., Vertinsky, I. & Krebs, J. R. (1974). The survival value of flocking in birds: a simulation model. *J. anim. Ecol.* **43:** 785–820.

Todd, S. (1979). *Waterfowl: ducks, geese and swans of the world.* Sea World Press, San Diego.

Ward, P. & Zahavi, A. (1973). The importance of certain assemblages of birds as 'information-centres' for food finding. *Ibis* **115:** 517–34.

Whitlock, R. J. (1979). The eco-physiology of certain wader species. BSc thesis, Stirling University, Stirling, UK.

Ydenberg, R. C. & Prins, H. H. Th. (1981). Spring grazing and the manipulation of food quality by barnacle geese. *J. appl. Ecol.* **18:** 443–53.

Ydenberg, R. C., Prins, H. H. Th. & van Dijk, J. (1983). Post-roost gatherings of barnacle geese: information centres? *Ardea* **71:** 125–32.

Yom-Tov, Y., Imber, A. & Otterman, J. (1977). The microclimate of winter roosts of the starling *Sturnus vulgaris* in Israel. *Ibis.* **119:** 366–8.

8

The unsociable plover – use of intertidal areas by Grey Plovers

D. J. TOWNSHEND, P. J. DUGAN &
M. W. PIENKOWSKI

Introduction

Grey Plovers (*Pluvialis squatarola*) breed on tundra in Siberia
and North America but, like most other arctic-breeding shorebirds, they
spend most of the year in wetlands further south, the Siberian population wintering from Britain and the Dutch Wadden Sea southwards to
much of Africa. In these areas they feed on intertidal flats and coasts,
standing and waiting for prey animals to reveal themselves before
darting to catch them – a feeding method typical of plovers.

Fig. 8.1. The intertidal areas on the Tees estuary, 1975–81, with (inset)
the positions of Lindisfarne and Teesmouth, the two study areas
mentioned in the text.

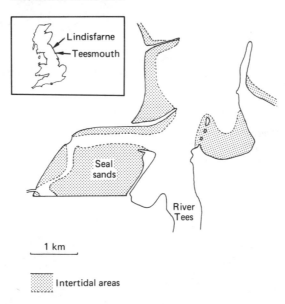

Lindisfarne

Teesmouth

Seal
sands

River
Tees

1 km

Intertidal areas

On the Tees estuary in northeast England, up to 350 Grey Plovers may be present. We began detailed observations there in 1975, but information on the population has been collected since January 1973. Our main study area was Seal Sands (Fig. 8.1), a nutrient-rich area of mudflats (140 ha) bounded on three sides by reclamation walls, which facilitated viewing of the birds.

The number of Grey Plovers on the estuary increased gradually from July to October and usually fell again, before increasing rapidly once or twice in mid-winter, to reach a peak in February or March (Fig. 8.2). (Seven records of colour-ringed birds show that Grey Plovers from the Tees use the coasts of Denmark, East and West Germany and the

Fig. 8.2. Seasonal changes in the total population of Grey Plovers on the Tees estuary, and in the numbers of flock-feeding and long-term territorial birds on Seal Sands. (*a*) 1978–79; (*b*) 1979–80. Open circles, total; squares, flock; solid circles, long-term territorial.

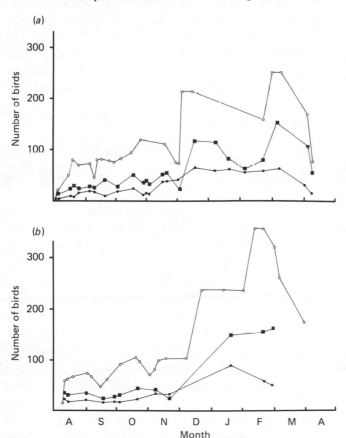

Netherlands in spring and autumn.) The birds fed on the intertidal flats when these were exposed. Consequently they moved, to some extent, with the tide. However, they occurred mainly on the drier intertidal mudflats, avoiding softer muds more densely populated by other shorebirds. Yet it was clear from watching the Grey Plovers on these mudflats that, although they roosted together, they did not all forage in the same manner.

In order to investigate the differences in feeding behaviour between individual Grey Plovers, we marked over 200 with unique combinations of coloured leg-rings. Watching these birds showed that, for most of the time the mud was exposed, each individual stayed in a limited part of Seal Sands. Some defended feeding territories in the same place, day and night, for many days, sometimes throughout the winter, and even in successive winters. Others did not engage in territorial defence, and sometimes changed their feeding sites from day to day. Still other individuals defended territories on some occasions but not on others. Later in our study we discovered that some Grey Plovers behaved differently by night than by day.

In the first part of this chapter, we describe these uses of space in more detail and explore why they occur. Following this, we discuss how relevant these observations at Teesmouth may be to other sites.

Use of space in daytime

Patterns of use

There were two principal ways in which Grey Plovers used Seal Sands.

Some individuals defended a fixed feeding site against conspecifics over long periods of time (months). Once established, the site was usually defended until the bird left the estuary in spring. However, on some occasions, particularly in autumn, periods of intense territorial aggression were followed by changes in territory ownership, suggesting that some displacement occurred (see Dugan 1981*a* for further details).

Other individuals did not defend long-term territories, but fed without a territory for all or most of the winter. These birds, through avoidance and occasional aggression, were widely dispersed over the feeding grounds. However, sometimes these individuals did defend a fixed feeding site for a short period (several hours, days or occasionally weeks).

The social behaviour varied between different areas of Seal Sands and was related principally to certain physical characteristics of the mudflats

(Fig. 8.3). (1) Topography. A characteristic feature of most long-term territories was the presence of deep creeks (Dugan 1981*a*). In contrast, non-territorial birds and those occupying short-term territories fed where there were no deep creeks. Since Grey Plovers rely almost entirely on the visual detection of prey at the surface of the mud (Pienkowski 1983*a*), buffeting by strong winds can severely reduce their ability to detect their food. Grey Plovers on the exposed areas of Seal Sands stopped feeding at wind speeds in excess of 25–30 knots (12–14.5 m/s) (Dugan *et al.* 1981). But in deep creeks the wind speed was much reduced (Fig. 8.4) and the Grey Plovers were able to continue feeding even during the worst gales. Thus, the possession of a territory containing a creek enabled birds to feed at times in winter when this was impossible for birds feeding on the more exposed flats. (2) Tidal level and substrate type. Away from the areas where long-term territories were held, Grey Plovers sometimes occupied territories for short periods. Short-term territories were situated particularly on the higher, sandier flats which, of course, drained quickly, and so dried out relatively rapidly after exposure. This drying probably decreased the availability of the main prey at Seal Sands, *Nereis diversicolor* (Townshend 1981). In contrast, the areas at lower tidal levels and those with wetter substrates dried out slowly, so the decline in prey availability during the much shorter period of exposure was small. The density of Grey Plovers in these areas was, in general, high and short-term defence of space was rare.

Factors other than these physical characteristics may also have influenced social behaviour. In particular, we think that interspecific effects may have been important. The highest densities of feeding shorebirds occurred at the lower tidal levels. Since Grey Plovers did not, in general, exclude other species from their feeding sites, the absence of territoriality in these lower areas may have been, in part, the result of a high degree of interspecific competition for prey and of interference in searching.

Costs and benefits of territoriality

As a means of understanding when, and under what conditions, birds should adopt a particular spacing behaviour, several authors (e.g. Davies 1976; Myers, Connors & Pitelka 1979) have tried to weigh up the respective advantages and disadvantages of each alternative, and to determine the factors which affect the balance between them. This cost–benefit approach is applied here to the use of territories by Grey Plovers on the Tees. We believe that the time period over which the costs and benefits of a behaviour should be evaluated differs between the two principal categories of use of space. For long-term territory holders, the benefits must exceed the costs when considered over the whole time period for which the territory is held, even though, on a day-to-day basis, costs may sometimes exceed benefits. In contrast,

Fig. 8.3. The types of social behaviour shown by Grey Plovers on different areas of Seal Sands in relation to topography, substrate type and tidal level. (*a*) Substrate types and tidal level; (*b*) social behaviour in relation to areas; (*c*) long-term territories (hatched) in relation to topography; (*d*) long-term territories (hatched) in relation to substrate and tidal level.

(*a*)

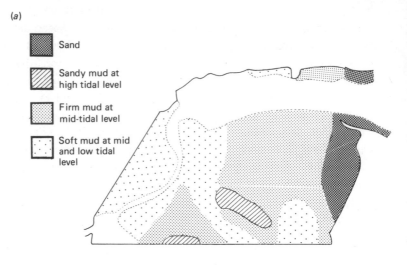

(*b*)

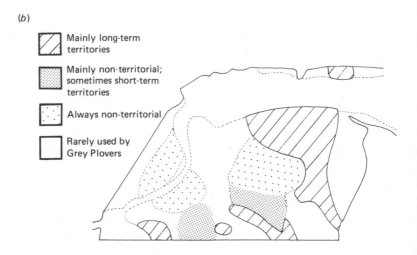

short-term territory holders probably assess the costs and benefits of site defence for each tidal cycle.

Long-term territoriality. The number of plovers defending long-term territories increased in November, whilst the number not doing so fell (Fig. 8.2). This increase was due to territories being established not only by newly arrived birds but also by birds that had previously been non-territorial (Dugan 1981*a*). Presumably the balance between costs and benefits for the latter individuals changed during the autumn.

Dugan (1981*a*) has suggested two major costs of territoriality at Teesmouth: (i) aggression associated with territorial defence, and (ii) an increased risk of predation, particularly by foxes (*Vulpes vulpes*) at night. Of these, only the former is known to change seasonally. Time spent in aggression by individually marked territorial birds was highest

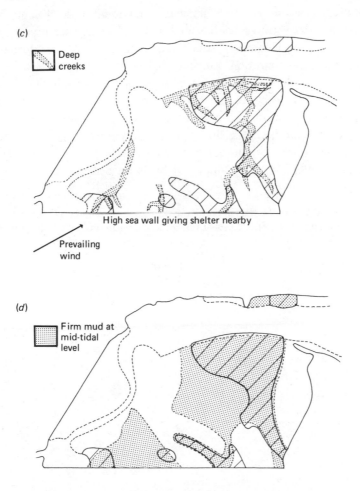

(c)

Deep creeks

High sea wall giving shelter nearby

Prevailing wind

(d)

Firm mud at mid-tidal level

in autumn and lowest in winter (P. J. Dugan, unpublished). There is no information on seasonal changes in predation.

Dugan (1981*a*) identified two major benefits of holding a long-term territory: (i) a reduction in the rate of depletion of prey, by excluding conspecifics and possibly by the selective use of patches of prey by the owner, and (ii) shelter, provided by the deep creeks, allowing the owners to forage during gales. Both benefits were of maximum value when conditions became most severe, and difficulties in maintaining a net positive energy balance were most likely to be experienced. The risk of severe weather rises markedly in northeast England in November, so the parallel increase in the number of territories established could be due to the increase in the value of these benefits at that time (Dugan 1981*a*) and to the simultaneous decline in the costs of territorial defence.

This discussion has assumed that each individual on Seal Sands was free to choose the strategy, whether territorial or non-territorial, that maximized its own chances of survival over the winter. An alternative explanation of the dimorphism in spacing behaviour is that all the areas suitable for one of these options were full, so that some individuals were forced to adopt the other option and so had no choice. On balance, the first alternative seems the more likely. For example, some Grey Plovers arriving in mid-winter took up territories on previously undefended areas. Furthermore, the number of birds using the non-territorial areas varied in parallel with the number of birds on Seal Sands as a whole

Fig. 8.4. Wind speeds in creeks within territories and on the nearby open mudflats. Speed measured on two days in territory 1, and on one day in the others.

Territory	Wind speeds: (knots)		Significance of difference
	On exposed flats	In deep creeks	
1	9.5	6.7	$P < 0.001$
	8.6	3.8	$P < 0.001$
2	8.7	3.3	$P < 0.001$
3	6.5	6.0	NS
4	7.9	4.8	$P < 0.001$

(Fig. 8.2), suggesting these areas were not full. However, some of the territorial areas may have been filled completely before November, since the eviction of long-term territory holders by newly arrived birds was observed occasionally (Dugan 1981a; Townshend 1981).

Short-term territoriality. Short-term defence of a feeding site occurred particularly in October and in mid-winter (from late December to February) (Townshend 1981). It thus coincided with the peaks in the population size (Fig. 8.2), and therefore the density, of Grey Plovers on Seal Sands, and with the coldest period of the winter.

Grey Plovers can capture only those prey that are at the surface of the substrate, but only a small proportion of the prey in the mud is to be found at the surface at any one time. By successfully excluding conspecifics from its feeding site, a Grey Plover may reduce the disturbance of its available prey, which would otherwise move down their burrows out of the bird's reach (cf. Goss-Custard 1970). Also, a territory holder is the only Grey Plover removing prey from its site so presumably the depletion of the available fraction is kept to a minimum. These benefits are likely to be greatest at low temperatures, when energy requirements are higher but prey availability lower than usual (e.g. Evans 1976; Pienkowski 1981a). Furthermore, we suggest that, by excluding conspecifics from its feeding site, a territory holder does not need to adjust its search path to avoid other feeding Grey Plovers. This benefit will be most important when Grey Plover density is high. Thus the two benefits reach a maximum at the times when short-term territoriality is most frequently observed, i.e. when bird density is high and temperatures low.

The proposed immediate advantages of improved prey availability and reduced interference in searching, whilst also accruing to long-term territory holders, are likely to be of much greater importance for the holders of short-term territories. As we have shown, short-term territories occurred mainly on higher, sandier flats. On lower flats, even in mid-winter, when the benefits of short-term territoriality were probably greatest, territorial defence was rare and of very short duration (for part of a single period of tidal exposure). This suggests that the energy expended in excluding conspecifics normally outweighed the resulting improvement in intake rate. On the higher, drier substrates in mid-winter, the converse was probably true, with the benefits outweighing the costs at certain times. Higher, sandier substrates would have cooled more rapidly than the lower wetter areas (because air and sand change temperature more quickly than does water). Consequently prey availability on the former areas should have fallen faster, thereby favouring territorial behaviour. It is possible that the birds feeding on the higher, drier areas were subordinate individuals, unable to feed adequately in the main non-territorial group on the lower areas. These subordinates might then have to defend their feeding sites in order to maintain an adequate intake rate when temperatures were low. This is the subject of continuing research.

Individual differences in the use of space

The main dichotomy in the use of space was for birds either to defend a long-term territory or to feed non-territorially for most or all of the winter. Some of the Grey Plovers arriving at Teesmouth established long-term territories immediately, even as early as August, whilst others established them later in the year. In contrast, some birds that arrived early in autumn fed non-territorially throughout the winter. For these individuals, the net advantages of non-territoriality may have continued to outweigh those of territoriality. These individual differences persisted because, from their second winter onwards, most individuals showed the same use of space in successive years (Dugan 1981a; Townshend 1981).

The reasons for these individual differences are unknown, though there are several possibilities. Although the birds which established long-term territories did not appear to be bigger than the others, they included a smaller proportion of juveniles, and may have been more aggressive, more experienced or more skilled feeders, and therefore capable of obtaining the maximum benefit from the territory at the least cost in defence. Conversely, the non-territorial birds may have been the more aggressive or more skilled birds and were thus able to maintain a positive energy balance without the need to exclude conspecifics from their feeding sites. The non-territorial group may even comprise a mixture of the latter together with much poorer individuals unable to acquire a territory. Another possibility is that the difference between individuals reflects a behavioural polymorphism. The long-term benefits of territoriality (particularly the conservation of food stocks) are likely to be greatest in severe weather, leading to higher survival amongst territorial than non-territorial birds. The advantage of this behaviour in hard winters may be offset, in mild winters, by the higher survival of non-territorial birds which do not incur the costs of defence. Over a period of good and bad years, then, the two strategies may contribute equally to fitness so that both are maintained.

Winter territoriality and regulation of density

Although territoriality must be selected for at the individual level, it can have consequences at the level of the population. By preventing other Grey Plovers from settling in an area, territorial behaviour could play a role in regulation of density during the winter, analogous to the limiting effect of territories upon the breeding densities of many birds, e.g. Red Grouse (*Lagopus l. scoticus*, Watson 1967), Tawny Owls (*Strix aluco*, Southern 1970). Great Tits (*Parus major*, Krebs 1971). Alternatively, territorial behaviour might merely result in the spacing out of individuals, but impose no limits on their density (Fretwell 1972; Davies 1978).

A few areas of Seal Sands were apparently filled by territorial birds in autumn, and then held a more or less constant number through the winter (Townshend 1981, and see above). As all the birds there were territorial and continued to drive out intruders throughout the winter, territorial behaviour must have limited, to some extent, the number settling on these areas. It was not simply a method of spreading the individuals over the mudflats.

However, territoriality was absent in several areas so that birds without long-term territories were able to feed elsewhere within Seal Sands throughout the winter. Consequently, territorial behaviour by itself could not have limited the population of Seal Sands as a whole. But the population could be limited if, in addition to territorial behaviour on some areas, there was some mechanism of density regulation in the non-territorial areas. Here density might be limited through a dominance hierarchy, in which the lowest individuals were unable to feed at a sufficient rate, so left (cf. Goss-Custard & Durell, Chapter 11 of this volume). However, at present there is no evidence for such a mechanism in Grey Plovers.

Behaviour at night

In the preceding sections we have discussed the possible advantages and disadvantages of the different patterns of spacing behaviour shown by Grey Plovers in the daytime. Many shorebirds, including Grey Plovers, feed by night as well as by day. By attaching miniature radio transmitters to 26 Grey Plovers, we found that, on the Tees estuary, the majority was present on the feeding grounds at night. If their nocturnal foraging differs from that during the day, the balance of costs and benefits over 24 hours may be greatly affected. Indeed, the balance may have to be considered over longer periods, since the success of nocturnal foraging may be affected by the spring/neap tidal pattern, as discussed below.

At night, availability of some prey species is higher (Dugan 1981*b*). In some areas exposed only at low water on spring tides the very large polychaete *Nereis virens*, which does not come to the surface during the day, is present on the surface at night. We suspect that a plover feeding on *N.virens* could take in food at a much faster rate than if feeding on the much smaller *N. diversicolor*. If true, it would be advantageous for Grey Plovers to move to areas with high densities of *N. virens* when these are exposed on night-time spring tides. However, if one of the major benefits of long-term territoriality is indeed the conservation of food stocks as we have argued, a territory holder should remain in its

Fig. 8.5. A comparison of day (unshaded) and night (stippled) usage of feeding sites at Teesmouth, January–April 1981, by three behavioural categories of marked Grey Plovers. (*a*) Long-term territory holders; (*b*) non-territorial bird; (*c*) short-term territory holders.

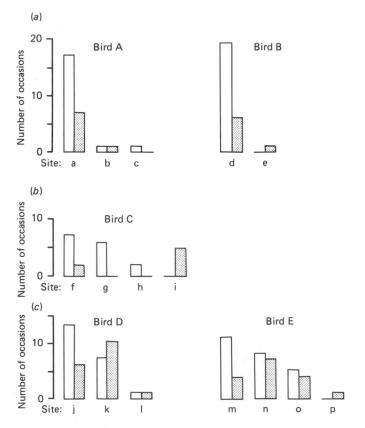

territory to exclude others whenever the feeding conditions there are good enough to attract other plovers, by both day and night. Territory holders should often be unable to take advantage of the areas rich in *N. virens* at night. In contrast, non-territorial birds are not constrained, and so should be able to move to more profitable areas at any time.

Preliminary results from Grey Plovers carrying radio transmitters confirm this distinction between long-term territory holders and non-territorial birds. Individuals defending long-term territories were generally present there both night and day (e.g. Fig. 8.5*a*). (We do not know as yet whether they defend their sites at night.) In contrast, at least some non-territorial Grey Plovers frequently fed in sites at night that were not used during the day (e.g. Fig. 8.5*b*). At night, the non-territorial birds fed at lower tidal levels (where *N. virens* is found) on spring tides rather than on other tides (Fig. 8.6). This was not so for long-term territory holders.

The same distinction would also be expected between birds holding long-term and short-term territories. We argued earlier that short-term territoriality provides the immediate benefit of increased prey availability. Short-term territory holders can, therefore, abandon their territory and feed elsewhere without loss of benefit from their territorial behaviour. In contrast, long-term territory holders would lose their main benefits. Grey Plovers defending short-term territories were, as expected, less faithful by day and, like non-territorial birds, sometimes changed feeding sites at night (Fig. 8.5*c*), by moving to low tidal areas on spring tides (Fig. 8.6).

If these differences in feeding behaviour between individuals at night are widespread in the Tees population – and we suspect that this is so – what are the implications? If it is true that the main prey species are more accessible at night and that most of the Grey Plovers feed then, nocturnal foraging may provide a large proportion of the daily energy intake (Dugan 1981*b*), especially during the long nights in mid-winter. The day, rather than the night as is normally assumed for waders, may therefore provide the 'topping-up' feeding period for Grey Plovers on the Tees (see Chapter 2 of this volume).

The importance of night feeding may vary according to the way a bird uses space. Individuals defending long-term territories may have an approximately constant rate of intake through 24 hours, whereas that of other birds may be considerably greater at night than during the day (Fig. 8.7*a*). Amongst birds defending short-term territories or feeding non-territorially, the nocturnal feeding rate may be much higher on spring tides than on neaps (Fig. 8.7*b*). The importance of night feeding may vary seasonally also. It appears that, on the Tees estuary, *Nereis virens* may become active at the surface from late December to spring only (A. G. Wood, personal communication), so night feeding conditions for non-territorial and short-term territorial birds may improve just when energy demands are increasing. Thus, the benefits of each pattern of use of space may vary considerably during a single winter. An increase in the supply of available prey in late December could help to explain why the mid-winter influxes of Grey Plovers can be supported on the estuary.

Fig. 8.6. Night feeding sites of marked Grey Plovers at Teesmouth, January–April, 1981: a comparison between usage on spring tides (stippled) and other tides (unshaded). The tidal levels at which birds B and E defended territories are shown. (*a*) Long-term territory holder, bird B; (*b*) non-territorial bird, bird C; (*c*) short-term territory holder, bird E.

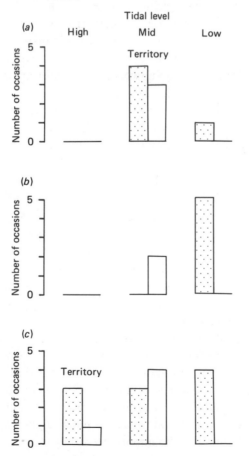

Comparison with other studies

These observations on individually marked Grey Plovers are restricted to one estuary, the Tees, and only to the years since 1975. However, this study can be compared with studies on another nearby coastal wetland, Lindisfarne (Fig. 8.1), from 1973 to 1976 (Pienkowski 1983*a*,*b*), and at Teesmouth in 1973 (Pienkowski unpublished), to see how widely the ideas might apply. In neither of these studies were Grey Plovers individually marked, but some information relevant to the present discussion is available.

Grey Plovers at Teesmouth 1973

A comparison between the two studies at Teesmouth is interesting because of the massive reclamation of Seal Sands that took place in the intervening period. The area of intertidal mudflats on Seal Sands

Fig. 8.7. Schematic representation of possible day-to-night variation in the rate of energy intake of Grey Plovers showing different patterns of use of space. (*a*) Comparison between long-term territory holders and other Grey Plovers; (*b*) comparison between the behaviours on spring and neap tides of birds not holding long-term territories.

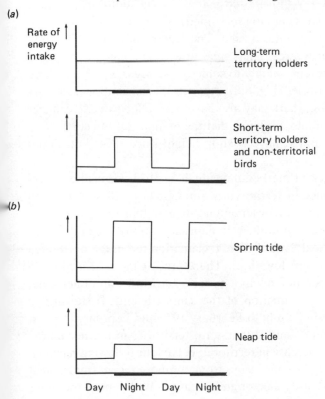

Table 8.1. *Mean numbers of Grey Plovers feeding in the north area of Seal Sands at low water in February*

		Mean number	Density (Grey Plovers/ha)
Pre-reclamation	1973	93	0.66
Post-reclamation	1976	118	0.84
	1977	177	1.26
	1978	174	1.24
	1979	133	0.95
	1980	213	1.52
	1981	163	1.16

was reduced by 60 % (Fig. 8.8) but the number of birds was not reduced in proportion. In fact, the overall density of Grey Plovers, and of most other shorebird species on the remaining part of Seal Sands, increased markedly (Table 8.1). The shorebirds then consumed over 90 % of the standing crop of macrofauna (Evans *et al.* 1979).

Not surprisingly, the distribution of shorebirds was different in the earlier years. The Grey Plovers fed mainly in the south area of Seal Sands (the mudflats subsequently reclaimed), some parts of which were usually uncovered at all states of tide. The feeding areas used by Grey Plovers on the north area (the area now remaining) were more restricted than at present, mainly to some of the areas which now hold many long-term territories (Fig. 8.8). Apart from distributional changes caused by reclamation, there may also have been effects due to changes in sediment. For example, we know that the mudflats of the north area, particularly Central Bank, have risen in height since 1974, but do not know by how much.

Because of the lack of marked individuals in 1973, we cannot assess directly the occurrence of territoriality amongst Grey Plovers prior to reclamation. However, our observations suggest that there may have been less territorial behaviour then. In February, up to 86 % of the Grey Plovers feeding on Seal Sands before reclamation fed in the north area on spring, but not neap, low tides. Therefore, at most 14 %, i.e. 35 birds, could have been defending long-term territories on either area, because this requires occupation of the same (defended) site during each period of exposure. In both February 1979 and February 1980, at least 50 Grey Plovers defended long-term territories in the north area alone (Fig. 8.2). An increase in territorial behaviour following reclamation accords well with the conclusions from our more recent studies: that a reduction in prey density and/or an increase in bird density promotes an increase in territorial behaviour.

Grey Plovers at Lindisfarne 1973–76

Grey Plover behaviour was studied on the sandflats at Holy Island Sands, Lindisfarne National Nature Reserve, from 1973 to 1976. In this site, only 130 km north of the Tees, the substrate is considerably sandier than on Seal Sands. As a result, the main potential prey species present, and consequently the diet of the birds studied (Pienkowski 1981a, 1982), were different. Though we observed many similarities in

Fig. 8.8. Grey Plover feeding areas on Seal Sands before and after recent reclamation.

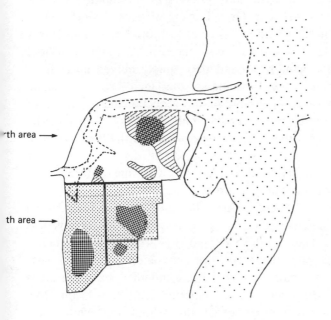

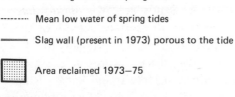

———— Mean high water of spring tides

-------- Mean low water of spring tides

═════ Slag wall (present in 1973) porous to the tide

▦ Area reclaimed 1973–75

▦ Main feeding areas of Grey Plovers before reclamation

▨ Main areas holding Grey Plover long-term territories after reclamation the behaviour of the Grey Plovers there to those on the Tees, there were also some marked differences.

Feeding rates at Lindisfarne decreased significantly with falling temperature and increasing wind force, rainfall and drying out of the substrate (Pienkowski 1983b), patterns which were reflected in general at Teesmouth. Observations of invertebrate behaviour at Lindisfarne revealed that these changes in feeding rate coincided with changes in surface activity of prey (Pienkowski 1981b), as postulated for daytime feeding at Teesmouth. At night, some prey species at Lindisfarne were more active than by day, and the Grey Plovers probably took these. However, in contrast to the behaviour observed at Teesmouth, very few aggressive interactions between Grey Plovers were observed at Holy Island Sands, and there was no evidence of territorial behaviour. Grey Plovers on Holy Island Sands were not spaced out through territorial behaviour during the winter; instead they simply moved away from others in adverse conditions. In doing so, they increased their search area, and this may have compensated for any reduction in the number of cues from prey caused by the adverse conditions (Pienkowski 1983a).

Why was there such a marked difference between the behaviours of Grey Plovers in the two study areas? Why did they defend territories at Teesmouth but not at Holy Island Sands? Can the difference be correlated with any features of the areas?

1. Topography and prey behaviour. The study area on Holy Island Sands comprised sandflats without deep gullies. Grey Plovers have less opportunity than at Teesmouth to defend sites which provide shelter during gales. (It is possible that territories were defended in other parts of Lindisfarne, e.g. Fenham Flats which are muddier, but such areas were not studied.) However, the reason why Grey Plovers on Holy Island Sands remained non-territorial throughout the winter was perhaps because they could continue feeding, albeit at a reduced rate, at wind strengths exceeding those at which foraging stops on the open flats at Teesmouth (Fig. 8.9). This may be because the prey most taken by Grey Plovers on Holy Island Sands are *Notomastus latericeus* and *Arenicola marina*. The cues from the surface activity of these animals are more obvious to a human observer, and probably to a Grey Plover, than those of *Nereis diversicolor*, the main prey at Teesmouth.

2. Interspecific competition. On Holy Island Sands, Ringed Plovers (*Charadrius hiaticula*), Bar-tailed Godwits (*Limosa lapponica*) and Dunlins (*Calidris alpina*) also take *Notomastus latericeus* from the same area as Grey Plovers, for at least part of the tidal cycle. These waders may, therefore, compete with Grey Plovers, and also cause disturbance

of prey and interference in searching. Therefore, the exclusion of conspecifics by a Grey Plover would yield only a small improvement in the density of available prey, because all the other shorebirds are also depleting the same prey stocks. Perhaps the absence of intraspecific territorial defence in Grey Plovers is partly a consequence of this. In contrast, at Teesmouth, the levels of interference and competition between Grey Plovers and other species are probably quite low. Although *Nereis diversicolor* is also the main food of Curlews (*Numenius arquata*) and Bar-tailed Godwits, most Grey Plover territories are situated in areas where few of these species feed. Where Curlews do

Fig. 8.9. Differences in response to wind of feeding Grey Plovers at Lindisfarne and Teesmouth. Although the rate of capture of worms at Holy Island Sands, Lindisfarne, during autumn and winter decreased significantly with increasing wind speed (regression line shown; $P<0.001$), most birds continued to feed in winds stronger than those associated with the cessation of feeding at Teesmouth (indicated by shading).

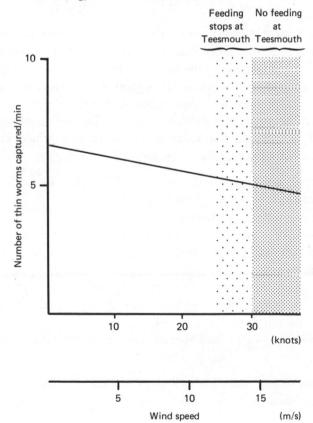

feed in Grey Plover territories, they also defend long-term territories (Townshend 1981). This may further reduce competition for food, and may enable the benefits of intraspecific territorial defence to outweigh the costs arising from interspecific competition.

3. Bird density. As noted earlier, the overall density of Grey Plovers, and of other waders, on Seal Sands is very high, perhaps partly because of the amount of reclamation that has taken place. Overall bird densities at Lindisfarne, and elsewhere, are much lower, although high concentrations may occur for brief periods on some areas. The territoriality on Seal Sands may, therefore, be a response to the high density of Grey Plovers there. However, territoriality has been recorded in wader species in several wintering areas, including some unaffected by reclamation (Myers *et al.* 1979), and signs of such behaviour in Grey Plovers have been observed by us at many sites. Thus it seems unlikely that this behaviour is restricted to artificially reduced habitats, though it may be enhanced in such areas.

Some predictions

Despite having a very stereotyped foraging behaviour (Pienkowski 1983*a*), it is clear that Grey Plovers show great variation, both within and between individuals, in the way that they utilize a feeding area. The results presented in this chapter suggest some factors which may influence the distribution and spacing behaviour of Grey Plovers within other intertidal areas. We therefore predict that the following factors will tend to increase the incidence of territorial behaviour (although they are not all prerequisites for it):

1. Unevenness and dissection by creeks of the feeding area;
2. Inconspicuousness of cues given by prey;
3. Rapid drying out of the substrate;
4. Low density of available prey (either low absolute density, or generally low availability because of, e.g. prevailing low temperatures);
5. Marked short-term depressions of prey availability and capture rate by, e.g. temperature, wind and tidal drying-out effects;
6. High densities of Grey Plovers; and
7. Low densities of other bird species feeding on the same prey species.

We would like to visit sites throughout Europe, Africa and other parts of the world to test these predictions, or encourage other workers to do so.

References

Davies, N. B. (1976). Food, flocking and territorial behaviour of the pied wagtail *Motacilla alba yarrellii* in winter. *J. Anim. Ecol.* **45**: 235–53.

Davies, N. B. (1978). Ecological questions about territorial behaviour. In *Behavioural ecology: an evolutionary approach*, ed. J. R. Krebs & N. B. Davies, pp. 317–50. Blackwell, Oxford.

Dugan, P. J. (1981*a*). Seasonal movements of shorebirds in relation to spacing behaviour and prey availability. Unpublished PhD thesis, University of Durham, South Road, Durham DH1 3LE, England.

Dugan, P. J. (1981*b*). The importance of nocturnal foraging in shorebirds: a consequence of increased invertebrate prey activity. In *Feeding and survival strategies of estuarine organisms*, ed. N. V. Jones & W. J. Wolff, pp. 251–60. Plenum Press, New York.

Dugan, P. J., Evans, P. R., Goodyer, L. R. & Davidson, N. C. (1981). Winter fat reserves in shorebirds: disturbance of regulated levels by severe weather conditions. *Ibis* **123**: 359–63.

Evans, P. R. (1976). Energy balance and optimal foraging strategies in shorebirds: some implications for their distributions and movements in the non-breeding season. *Ardea* **64**: 117–39.

Evans, P. R., Herdson, D. M., Knights, P. J. & Pienkowski, M. W. (1979). Short-term effects of reclamation of part of Seal Sands, Teesmouth, on wintering waders and Shelduck. I. Shorebird diets, invertebrate densities, and the impact of predation on the invertebrates. *Oecologia* **41**: 183–206.

Fretwell, S. D. (1972). *Populations in a seasonal environment*. Princeton University Press, Princeton.

Goss-Custard, J. D. (1970). Feeding dispersion in some overwintering wading birds. In *Social behaviour in birds and mammals*, ed. J. H. Crook, pp. 3–35. Academic Press, London.

Krebs, J. R. (1971). Territoriality and breeding density in the great tit, *Parus major* L. *Ecology* **52**: 2–22.

Myers, J. P., Connors, P. G. & Pitelka, F. A. (1979). Territoriality in non-breeding shorebirds. In *Shorebirds in marine environments*, ed. F. A. Pitelka, pp. 231–46. Cooper Ornithological Society.

Pienkowski, M. W. (1981*a*). Differences in habitat requirements and distribution patterns of plovers and sandpipers as investigated by studies of feeding behaviour. *Verh. orn. Gas. Bayern* **23**: 105–24.

Pienkowski, M. W. (1981*b*). How foraging plovers cope with environmental effects on invertebrate behaviour and availability. In *Feeding and survival strategies of estuarine organisms*, ed. N. V. Jones & W. J. Wolff, pp. 179–92. Plenum Press, New York.

Pienkowski, M. W. (1982). Diet and energy intake of Grey and Ringed Plovers, *Pluvialis squatarola* and *Charadrius hiaticula*, in the non-breeding season. *J. Zool., Lond.* **197**: 511–49.

Pienkowski, M. W. (1983*a*). Changes in the foraging pattern of plovers in relation to environmental factors. *Anim. Behav.* **31**: 244–64.

Pienkowski, M. W. (1983*b*). The effects of environmental conditions on feeding rates and prey-selection of shore plovers. *Ornis Scand.* **14**: 227–38.

Southern, H. N. (1970). The natural control of a population of Tawny Owls (*Strix aluco*). *J. Zool., Lond.* **162**: 197–285.

Townshend, D. J. (1981). The use of intertidal habitats by shorebird populations, with special reference to Grey Plover (*Pluvialis squatarola*) and Curlew (*Numenius arquata*). Unpublished PhD thesis, University of Durham, South Road, Durham DH1 3LE, England.

Watson, A. (1967). Population control by territorial behaviour in Red Grouse. *Nature* **215**: 1274–5.

9

Age-related distribution of Dunlin in the Dutch Wadden Sea

T. M. van der HAVE, E. NIEBOER &
G. C. BOERE

Introduction

Twice a year Dunlins (*Calidris alpina*) arrive in the Dutch Wadden Sea. In spring they come from their winter quarters, moult their body feathers (Boere 1976) and accumulate large fat reserves for the return migration to their breeding areas (Boere & Smit 1981). In autumn the picture is different because juveniles (birds in their first year) and adults behave differently. Juveniles depart later from the breeding areas (Soikkeli 1967), migrate over a larger area through Scandinavia (Mascher 1966; Folkestad 1975; Hardy & Minton 1980) and converge along the Norwegian coast (Ogilvie 1963; Nørrevang 1955;

Fig. 9.1. Location of trapping stations in the Dutch Wadden Sea with the mean percentages (± 1 s.D.) of juveniles in samples collected during the period 1970–80. Nb is the number of birds in the sample and Ny the number of years samples were obtained.

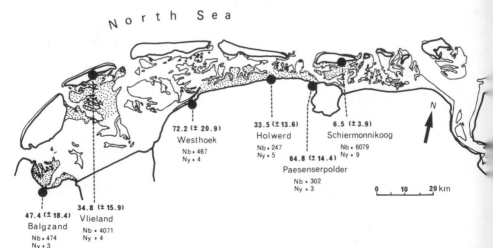

Leslie & Lessells 1978). By the time the juveniles reach the Dutch Wadden Sea, the adults have already arrived and are in an advanced stage of moult of both flight and body feathers (Nieboer 1972; Boere, de Bruijne & Nieboer 1973; Boere 1976).

Despite the detailed knowledge of the migration of this species, no studies have been made up to now on differences in the distribution of adult and young Dunlin. Such differences have been found in other species within one tidal area (Prater 1979; Atkinson *et al* 1981; Chapter 11 of this volume), but no study has been done on any species in an area the size of the Dutch Wadden Sea. This chapter analyses the distribution of adult and juvenile Dunlin in the Dutch Wadden Sea from variations in the numbers of birds and their age ratios determined in six different localities (Fig. 9.1). The migration routes through Scandinavia and the Baltic are also analysed in case Dunlins arriving in different parts of the Dutch Wadden Sea come from different breeding areas. The results are discussed in relation to the theory of habitat distribution of Fretwell & Lucas (1970).

Methods

The age ratios are based on mist-net samples. These are known to be biased in favour of juveniles in Oystercatchers (Goss-Custard *et al* 1981), and probably other species too (Pienkowski & Dick 1976); so they may only be used for comparisons between different localities. Densities on the foraging grounds were calculated by dividing the numbers counted at the adjacent high-tide roosts by the area of shore exposed at low water (cf. Goss-Custard, Kay & Blindell 1977; areas from Boere 1973). This is rather a rough method, but counts at low tide are difficult to carry out over such huge intertidal areas. The data used come from several published and unpublished sources (Boere 1973; Boere & Zegers 1974, 1975, 1977; Smit 1977; P. M. Zegers, unpublished). For statistical tests, juvenile percentages were transformed to angles with the arcsine transformation to normalize the data (Sokal & Rohlf 1969).

Results

Age ratios in different parts of the Dutch Wadden Sea

There were large differences in the age structure of autumn populations moulting and staging in different parts of the Dutch Wadden Sea but no geographical trend is apparent (Fig. 9.1). Although the variances were similar (Bartlett's chi-square: $0.5 > P > 0.1$), significance tests on differences between means were not carried out because variations in the percentages were not entirely independent of each other (see below).

The differences in age structure were consistent over many years (Fig. 9.2), despite the large fluctuations attributable to the considerable annual variation in breeding success (Holmes 1966; Soikkeli 1967) which is typical of arctic birds in general. Fig. 9.2 shows that juvenile percentages fluctuated in synchrony in all areas but that the fluctuations were least on Schiermonnikoog, which had the smallest proportion of young birds overall.

The migration of juveniles later than the adults could have affected the age ratios during autumn. Fig. 9.3 gives a rough comparison of the changes in age ratios in all areas. It shows that the percentage of juveniles increased everywhere during the course of the season. It also confirms that the percentage of juveniles was lowest on Schiermonni-koog but indicates that the difference may have increased during the season. A more detailed comparison is therefore given in Fig. 9.4, where the mean juvenile percentages are given for 10-day periods on Vlieland and Schiermonnikoog. Since waders were trapped on these islands simultaneously and frequently during the period 1971–75 and the numbers of Dunlin obtained were relatively large, these data may provide a reliable picture for the Dutch Wadden Sea as a whole. They suggest that (i) the percentage of juveniles increased considerably on both islands in the second and third 10-day periods of August, (ii) in the

Fig. 9.2. Annual percentages of juveniles during autumn in six localities of the Dutch Wadden Sea: Schi (Schiermonnikoog), V(Vlieland), H(Holwerd), B(Balgzand), P(Paesenserpolder) and W(Westhoek).

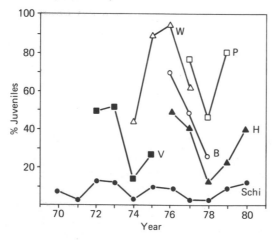

first two 10-day periods of September, there was only a small increase in the percentage of juveniles on Vlieland and even a decrease on Schiermonnikoog, and (iii) the percentage of juveniles increased considerably up to the last 10-day period of October: some decrease occurred thereafter but the percentage remained higher on both islands in comparison with August and the first part of September.

The second conclusion here suggests that juveniles arrived in two 'waves', one before and one after the first week of September, and that this effect was most pronounced on Schiermonnikoog. This is supported by the recapture rate of juveniles before and after 10 September. Of the 342 birds ringed during the study on Vlieland and Schiermonnikoog between 1 July and 10 September, only one (0.3%) was recaptured between September and 30 November. In contrast, 1.7% of the 1365 birds ringed after 10 September were recaptured before the end of November.

Age ratios and numbers of Dunlin

This analysis was based on data from Boere & Smit (1981) and a series of other counts made in several parts of the Dutch Wadden Sea. The Dutch Wadden Sea was divided into two areas, A and B (Fig. 9.5),

Fig. 9.3. Mean percentages of juveniles per month over several years in the period 1970–80. The data of Westhoek and Paesenserpolder were taken together.

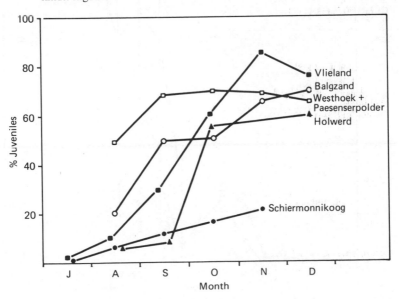

of approximately the same size. They differed in the season when maximum numbers of Dunlin occurred there. Area A includes localities where peak numbers were present in August and September (Schiermonnikoog, Terschelling and Ameland) whereas maximum numbers occurred in area B (Balgzand, Frisian coast, Texel and Vlieland) in September or later. Age ratios for area B were obtained by combining the results from several trapping areas, while those for area A were based on catches from Schiermonnikoog only.

From Fig. 9.5 it can be seen that: (i) the highest numbers of Dunlin were present in area A in August and September and in area B in

Fig. 9.4. Mean percentages of juveniles per 10-day periods on Schiermonnikoog and Vlieland over the period 1971–75. Each vertical line indicates ±1 s.e. The total numbers trapped are: 3934 (Vlieland) and 3025 (Schiermonnikoog).

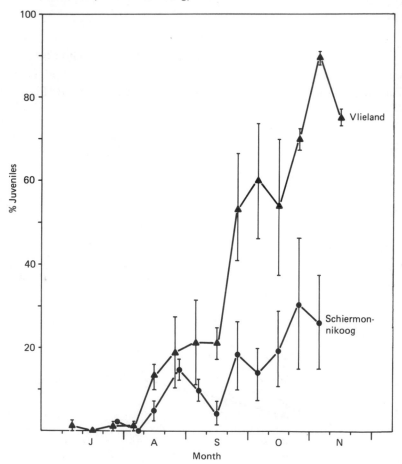

September and October. In area A the numbers decreased after September, in area B after October; (ii) the highest numbers of adults were present in August and September in both regions and the numbers began to decrease during September; (iii) the highest numbers of juveniles were present in October in both regions. Thereafter their proportion continued to increase, though their absolute numbers decreased in both areas, and (iv) the numbers of Dunlin were about equal in both areas in November, but the difference in the proportion of juveniles remained.

Age ratio and density of Dunlin
Because the size of the foraging grounds in areas A and B were about equal, Fig. 9.5 suggests that most juveniles occurred where Dunlin densities were low. To test this, the percentage of juveniles in mist-net samples was plotted against the density of Dunlin on the feeding areas calculated from roost counts made in the same area in the same month (Fig. 9.6). As expected, there was a significant negative correlation between juvenile percentage and Dunlin density ($P<0.01$). It is possible that this correlation was partly the result of the late arrival of juveniles and early departure of adults. However, the analysis of data from different periods separately showed that the negative correlations persisted and were mostly significant despite the small number of

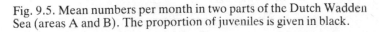

Fig. 9.5. Mean numbers per month in two parts of the Dutch Wadden Sea (areas A and B). The proportion of juveniles is given in black.

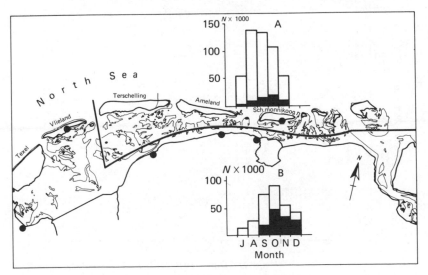

observations available: August, $r = -0.60$, $P < 0.10$ and > 0.05, $n = 9$; September, $r = -0.73$, $P < 0.05$, $n = 10$; October–December, $r = -0.68$, $P < 0.05$, $n = 12$.

Migration route

Data from birds ringed in Scandinavia and the Baltic and recaptured on Schiermonnikoog and Vlieland during the period 1971–75 were analysed in order to discover the migration routes of birds using different parts of the Dutch Wadden Sea. The ringing sites are shown in Fig. 9.7a,b, and the insets summarize the direction of migration from ringing site to the Dutch Wadden Sea. It is assumed here that Dunlins migrated in straight lines, an assumption made plausible by several studies in Scandinavia and Great Britain (Evans 1968; Mascher 1971; Folkestad 1975). Scandinavia was divided into five sectors of 18° with the Dutch Wadden Sea as angular point and all the directions of flights from the ringing sites distributed between them. The results showed that none of the recoveries on either Vlieland or Schiermonnikoog were of birds ringed for the first time as adults in Norway and the majority of those ringed as juveniles were caught in Norway, although there were

Fig. 9.6. The relationship between the percentage of juveniles in flocks and the density of Dunlin on the feeding grounds in three localities in the Dutch Wadden Sea. Squares, Balgzand; triangles, Vlieland; circles, Schiermonnikoog.

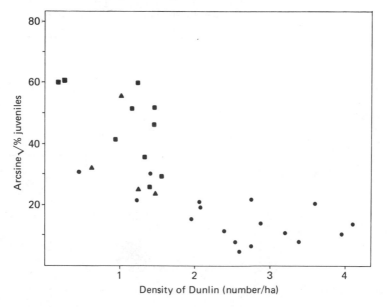

some records from the Baltic area. This confirmed that juveniles migrated over a wider area than adults through Scandinavia and the Baltic, but showed that there were no large differences in the migration route of the two islands' populations. In this respect, the different areas in the Dutch Wadden Sea were comparable.

Age-related distribution: an hypothesis

We suggest that the distribution of Dunlins in the Dutch Wadden Sea can be understood in terms of an overflow system. The island areas are occupied first, mainly by adults, and the highest numbers are reached at the end of August and the beginning of September. The coastal areas reach their maximum numbers in September and October, with proportionately and absolutely higher numbers of juveniles occurring there. This suggests that the young birds arriving later are unable to settle around the islands because of the large numbers of adults already present, and so they move on to less favoured areas inshore where bird density is lower.

In support of this hypothesis, the recaptures of juveniles indicate that the staging period of the first group of juveniles to arrive was short, and that this may have been related to the high densities of Dunlin present at the end of August and beginning of September. Moreover, this was most pronounced on Schiermonnikoog where higher densities occurred than on Vlieland. Furthermore, in years with a high breeding success when juveniles were abundant, they made more use of low density areas compared with years when they were scarce. From this it follows that the juvenile percentages in low density areas should have fluctuated more than in areas with high densities of Dunlin, and Fig. 9.2 demonstrates that this was so.

Discussion

Migration route

Like many birds, Dunlin show a high fidelity to their birth and breeding grounds (Soikkeli 1970), and there is evidence that they are also faithful to their moulting places (G. C. Boere and E. Nieboer, unpublished) and winter quarters (Pienkowski 1976). Apparently this site fidelity strengthens with age, because many waders migrate over a wider area in their first year (e.g. Jehl 1979) and Dunlins winter further south when they are young (Pienkowski & Dick 1975). Consequently, moulting populations of Dunlin with few juveniles must receive an influx of birds that moulted elsewhere in preceding years. This hypothesis is confirmed by ringing data. On Schiermonnikoog, where the juvenile percentage was low, there were more recaptures of birds that moulted in preceeding years in England than on Vlieland where the juvenile percentage was high (Table 9.1).

During autumn, juvenile Dunlins are frequently observed on beaches, saltmarshes and near freshwater ponds. Although this may be explained by a difference in habitat selection compared with adults (e.g.

Fig. 9.7. The Scandinavian and Baltic ringing sites of Dunlins recaptured on Vlieland (*a*) and Schiermonnikoog (*b*) during the period 1971–75. The insets show the frequency distribution of the directions of the ringing sites over five sectors between north and east with the Dutch Wadden Sea as angular point (solid symbols, ringed as juvenile; open symbols, ringed as adult. For other symbols see legend).

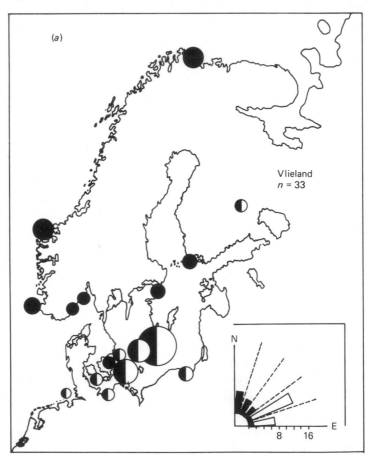

Bannerman 1961), it seems better in the light of our results to consider these habitats as marginal, low-density areas for Dunlins.

Habitat distribution

The increasing use of less preferred areas as numbers increase has been shown in several recent studies, but only on a small scale within one estuary: Redshank on the Ythan estuary (Goss-Custard 1977*a*), Oystercatchers and Knot on the Wash (Goss-Custard 1977*b*), Oystercatchers on the Exe estuary (Chapter 11 of this volume) and in

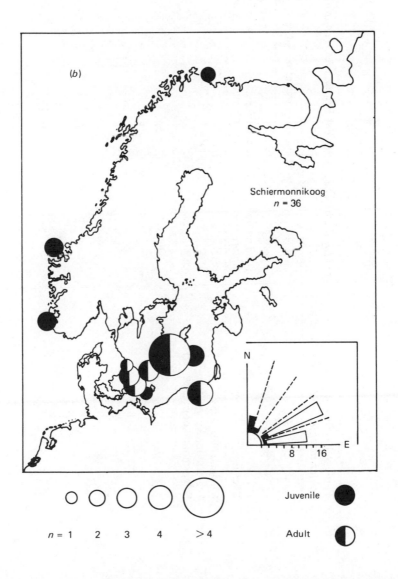

Table 9.1. *The numbers of Dunlins ringed for the first time in Great Britain and recaptured moulting on Schiermonnikoog and Vlieland*

Locality	Total number captured	Number of recaptures of	
		Dunlin ringed in Great Britain	Moulting Dunlin ringed in preceding years in Great Britain in August–September
Vlieland (1971–75)	3934	20	3
Schiermonnikoog (1971–75)	3025	16	6
Schiermonnikoog (1969–80)	6599	42	12

the Netherlands (Zwarts & Drent 1981) and waders and Teal on the Ventjagersplaat (Zwarts 1974, 1976). The distribution of Dunlin in the Dutch Wadden Sea also seems to be the result of an overflow system and provides a link between small-scale processes within an estuary and events occurring on a larger scale along migration routes. However, direct evidence of juvenile Dunlin ringed in a high-density area being retrapped a short time later in a low-density area is still lacking. This is probably because the trapping frequency was too low, or the duration of stay of juveniles in high-density areas too short, for such movements to be detected.

The theory of habitat distribution provided by Fretwell & Lucas (1970) suggests a mechanism which may account for the age-related distribution of Dunlin in the Dutch Wadden Sea, and of other waders elsewhere. The theory describes how organisms may distribute themselves over several habitats with differing suitabilities as population size increases. Depending on their type of social system, organisms will make more use of the marginal, less suitable habitats as population size rises because density-related effects increasingly depress the suitability of the intrinsically better areas. Some assumptions of this theory will now be discussed using the available evidence for waders in general and Dunlin in particular.

The first assumption is that the suitability of the habitat, in this case the feeding conditions, decreases as bird density increases. Possible mechanisms by which this could happen as far as waders are concerned have been reviewed by Goss-Custard (1980), where the increasing amount of evidence that interference does occur in the field is summarized. It seems that intake rate may often decrease in waders as their density rises.

The second assumption is that birds can estimate bird densities in different habitats. Although Fretwell & Lucas (1970) suggest that footprint abundance can serve as a cue for density and the mud in some tidal areas is an excellent substrate for this, more direct cues are available. The numbers of birds themselves are usually clearly visible in the open habitats and the behavioural consequences of increased density, whether through increased rates of encounter or reduced rates of food intake, are easy for observers to assess directly.

The third assumption is that individuals within one habitat will have identical success rates and are free to move from one habitat to another. This is derived from the belief that natural selection will favour individuals which maximize their success, however measured, so that all individuals will behave in the same, ideal way as they are all subjected to the same selection pressures. The evidence from waders, however, suggests that ideal behaviour of this kind may not often occur. For instance, intake rates may differ between areas of one estuary (Goss-Custard 1977a,b) and this would not occur if all birds were free to maximize their intake rate and to move to the combination of habitat quality and bird density which provides the highest intake rate available at the time. Similarly, the data for Dunlin from Vlieland and Schiermonnikoog suggest that juveniles may have had a higher annual mortality rate than adults, which may have been related to their lower weights in autumn and early winter (Fig. 9.8). Low weights are mainly caused by low fat reserves (Pienkowski, Lloyd & Minton 1979) and starvation seems to be the major cause of death during winter (Pilcher,

Fig. 9.8. Mean weights of juveniles (solid circles) and adults (open circles) ±1 s.e. (vertical lines) on Vlieland. The annual mortality rates are calculated with Haldane's (1955) method from recoveries of Dunlin ringed on Vlieland and Schiermonnikoog. Mortality rates based on recaptures in subsequent years are lower but show the same difference between age classes (T. M. van der Have, G. C. Boere and E. Nieboer, unpublished).

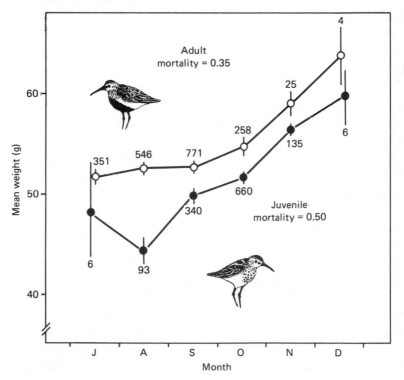

Beer & Cook 1974; Evans 1976), and predation is rare in the Dutch Wadden Sea (Boere 1976). Though not measured, these data may indicate that in autumn and winter, juveniles had a lower survival rate than adults, a difference already established for Oystercatchers on the Exe estuary (Goss-Custard & Durell, Chapter 11 of this volume) and in the Dutch Wadden Sea in severe weather conditions (Swennen, Chapter 10 of this volume). Therefore, the assumption that the behaviour of waders may be both ideal and free may not apply in these wintering waders. However, recent research on age-related differences in foraging behaviour in birds shows that immature, inexperienced individuals may forage less efficiently. This is most pronounced in fish-eating birds with a specialized foraging technique (e.g., Dunn 1972; Buckley & Buckley 1974), but has also been demonstrated in waders (Heppleston 1971; Groves 1978; Burger 1980). Obviously, juveniles need time to adapt to the tidal rhythm of a new locality and to learn to find the most profitable food patches (cf. Groves 1978). The main question in this context then is at what stage do juveniles become equal to adults in feeding experience so that differences in intake rates and survival can be assumed to be due to differences in competitive ability. We suspect that the 'learning period' is short in relation to the total period of stay in the Dutch Wadden Sea, and think that the lower survival rate of juveniles indicates dominance by adults over juveniles (cf. Fretwell 1969; Swingland 1977; Baker & Fox 1978). This is demonstrated in Turnstones (Groves 1978) and Oystercatchers (Goss-Custard & Durell, Chapter 11 of this volume) where adults dominated juveniles and the intake rate of juveniles was lower. If this is so, the 'ideal despotic' rather than 'ideal free' version of the Fretwell & Lucas (1970) model may be more appropriate as it assumes differences between individuals in their ability to compete for the most suitable feeding places.

If adult Dunlin are indeed dominant to juveniles, the black belly patch of adults could serve as a status signal (cf. Rohwer 1975; Rohwer & Rohwer 1978; Ulfstrand 1979). This patch is only present in the summer and gradually fades during the autumn moult. However, the moult of body feathers in Dunlin lags somewhat behind the primary moult (Boere 1976) so that during the major part of the period when the weights are low, the fat reserves minimal (G. C. Boere, H. H. Th. Prins and R. H. Drent, unpublished) and the wing surface strongly diminished by flight feather moult, the black belly patch is largely visible. A dominance signal may therefore be present during the period when adults presumably have higher energy demands than juveniles.

Clearly, only detailed behavioural and ecological research of the social structure of Dunlin flocks will enable us to interpret their age-related distribution in the Dutch Wadden Sea. Further studies will focus on the question of how the competitive abilities of juveniles develop and whether some form of status signalling is involved in the dominance relationships between adults and juveniles.

Acknowledgements

We would like to thank for providing data the following groups or persons: The Ringing Group Franeker (Westhoek), the Wader Ringing Group F.F.F. (Holwerd), Mr A. L. Pieters (Balgzand) and Mr P. M. Zegers (Paesenserpolder). We are particularly grateful to the numerous people who assisted in the fieldwork on Schiermonnikoog and Vlieland. We are very grateful to Mr W. Tamis for many fruitful discussions, Professor R. Drent and Dr G. Bryan for their comments on earlier drafts, Professor C. Wilkinson for correcting the English and Dr M. V. Hounsome for calculations of mortality rates. We thank Mr B. J. Speek, Vogeltrekstation, Arnhem, for his help to obtain permission for bird trapping. The figures were prepared by Mr G. W. H. van den Berg, the typing was done by Miss D. Hoonhout.

References

Atkinson, N. K., Summers, R. W., Nicoll, M. & Greenwood, J. J. D. (1981). Populations, movements and biometrics of the Purple Sandpiper *Calidris maritima* in eastern Scotland. *Ornis Scand.* **12:** 18–27.

Baker, M. C. & Fox, S. F. (1978). Dominance, survival, and enzyme polymorphism in dark-eyed juncos, *Junco hyemalis. Evolution* **32:** 697–711.

Bannerman, D. A. (1961). *Birds of the British Isles*, vol. 9. Oliver & Boyd, Edinburgh and London.

Boere, G. C. (1973). *Aantal, aantalsfluctuaties en verspreiding van 13 soorten steltlopers in het Nederlandse Waddengebied.* Intern Rapport Waddenzeecommissie, Leeuwarden. 89 pp.

Boere, G. C. (1976). The significance of the Dutch Waddenzee in the annual life cycle of arctic, subarctic and boreal waders. Part I. The function as a moulting area. *Ardea* **64:** 210–91.

Boere, G. C., de Bruijne, J. W. A. & Nieboer, E. (1973). Onderzoek naar de betekenis van het Nederlandse Waddengebied voor Bonte Strandlopers *Calidris alpina* in de nazomer en herfst. *Limosa* **46:** 205–12.

Boere, G. C. & Smit, C. J. (1981). Dunlin (*Calidris alpina* (L.)). In *Birds of the Wadden Sea,* ed. C. J. Smit & W. J. Wolff, pp. 157–69. Balkema, Rotterdam.

Boere, G. C. & Zegers, P. A. (1974). Wadvogeltelling in het Nederlandse Waddenzeegebied in juli 1972. *Limosa* **47:** 23–8.

Boere, G. C. & Zegers, P. M. (1975). Wadvogeltellingen in het Nederlandse Waddenzeegebied in april en september 1973. *Limosa* **48:** 74–81.

Boere, G. C. & Zegers, P. M. (1977). Wadvogeltellingen in het Nederlandse Waddenzeegebied in 1974 en 1975. *Watervogels* **2:** 161–73.

Buckley, F. G. & Buckley, P. A. (1974). Comparative feeding ecology of wintering adult and juvenile royal terns (Aves: Laridae, Sterninae). *Ecology* **55:** 1053–63.

Burger, J. (1980). Age differences in foraging Black-necked Stilts in Texas. *Auk* **97:** 633–6.

Dunn, E. K. (1972). Effect of age on the fishing ability of Sandwich terns *Sterna sandvicensis*. *Ibis* **114**: 360–6.

Evans, P. R. (1968). Autumn movements and orientation of waders in north-east England and southern Scotland, studied by radar. *Bird Study* **15**: 53–64.

Evans, P. R. (1976). Energy balance and optimal foraging strategies in shorebirds: some implications for their distributions and movements in the non-breeding season. *Ardea* **64**: 117–39.

Folkestad, A. O. (1975). Wetland bird migration in Central Norway. *Ornis Fenn.* **52**: 49–56.

Fretwell, S. (1969). Dominance behaviour and winter habitat distribution in juncos (*Junco hyemalis*). *Bird Banding* **40**: 1–25.

Fretwell, S. D. & Lucas, H. L., Jr (1970). On territorial behaviour and other factors influencing habitat distribution in birds, I. Theoretical development. *Acta Biotheor.* **19**: 16–36.

Goss-Custard, J. D. (1977*a*). Predator responses and prey mortality in redshank, *Tringa totanus*, and a preferred prey, *Corophium volutator*. *J. anim. Ecol.* **46**: 21–35.

Goss-Custard, J. D. (1977*b*). The ecology of the Wash. III. Density-related behaviour and the possible effects of a loss of feeding grounds on wading birds (Charadrii). *J. appl. Ecol.* **14**: 721–39.

Goss-Custard, J. D. (1980). Competition for food and interference among waders. *Ardea* **68**: 31–52.

Goss-Custard, J. D., Durell, S. E. A., Sitters, H. & Swinfen, R. (1981). Mist-nests catch more juvenile oystercatchers than adults. *Wader Study Group Bull.* **32**: 13.

Goss-Custard, J. D., Kay, D. G. & Blindell, R. M. (1977). The density of migratory and overwintering redshank, *Tringa totanus* (L.), and curlew, *Numenius arquata* (L.), in relation to the density of their prey in south-east England. *Est. coast. mar. Sci.* **5**: 497–510.

Groves, S. (1978). Age-related differences in Ruddy Turnstone foraging and aggressive behaviour. *Auk* **95**: 95–103.

Haldane, J. B. S. (1955). The calculation of mortality rates from ringing data. *Proc. 11th Int. Orn. Congr.* (Basel) 454–8.

Hardy, A. R. & Minton, C. D. T. (1980). Dunlin in Britain and Ireland. *Bird Study* **27**: 81–92.

Heppleston, P. B. (1971). The feeding ecology of Oystercatchers (*Haematopus ostralegus* L.) in winter in Northern Scotland. *J. anim. Ecol.* **40**: 651–72.

Holmes, R. T. (1966). Breeding ecology and annual cycle adaptations of the Red-backed Sandpiper (*Calidris alpina*) in Northern Alaska. *Condor* **68**: 3–46.

Jehl, J. R., Jr (1979). The autumnal migration of Baird's Sandpiper. In *Shorebirds in marine environments*, ed. F. A. Pitelka, pp. 55–68. Cooper Ornithological Society.

Leslie, R. & Lessells, C. M. (1978). The migration of Dunlin *Calidris alpina* through northern Scandinavia. *Ornis Scand.* **9**: 84–6.

Mascher, J. W. (1966). Weight variations in resting Dunlins (*Calidris alpina alpina*) on autumn migration in Sweden. *Bird Banding* **37**: 1–34.

Mascher, J. W. (1971). A plumage-painting study of autumn Dunlin *Calidris alpina* migration in the Baltic and North Sea areas. *Ornis Scand.* **2**: 27–33.

Nieboer, E. (1972). Preliminary notes on the primary moult in Dunlins *Calidris alpina*. *Ardea* **50**: 112–19.

Nørrevang, A. (1955). Rylens (*Calidris alpina* (L.)) traek i Nordeuropa. *Dansk. Orn. Foren. Tidsskr.* **49**: 18–49.

Ogilvie, M. A. (1963). The migrations of European Redshank and Dunlin. *Wildfowl Tr. Ann. Rep.* **14**: 142–9.

Pienkowski, M. W. (1976). Recurrence of waders on autumn migration at sites in Morocco. *Vogelwarte* **28**: 293–7.

Pienkowski, M. W. & Dick, W. J. A. (1975). The migration and wintering of Dunlin *Calidris alpina* in North-West Africa. *Ornis Scand.* **6**: 151–67.

Pienkowski, M. W. & Dick, W. J. A. (1976). Some biases in cannon- and mist-netted samples of wader populations. *Ringing & Migration* **1:** 105–7.

Pienkowski, M. W., Lloyd, C. S. & Minton, C. D. T. (1979). Seasonal and migrational weight changes in Dunlins. *Bird Study* **20:** 134–48.

Pilcher, R. E. M., Beer, J. V. & Cook, A. W. (1974). Ten years of intensive late-winter surveys for waterfowl corpses on the north-west shore of the Wash, England. *Wildfowl* **25:** 149–54.

Prater, A. J. (1979). Shorebird census studies in Britain. In *Shorebirds in marine environments*, ed. F. A. Pitelka, pp. 157–66. Cooper Ornithological Society.

Rohwer, S. (1975). The social significance of avian plumage variability. *Evolution* **29:** 593–610.

Rohwer, S. & Rohwer, F. C. (1978). Status signalling in Harris sparrows: experimental deceptions achieved. *Anim. Behav.* **26:** 1012–22.

Smit, C. J. (1977). *On the occurrence of 32 bird species in the Danish, German and Dutch Wadden Sea.* Unpublished report of the International Wadden Sea Working Group, part. 3. Rijksinstituut voor Natuurbeheer. 174 pp.

Soikkeli, M. (1967). Breeding cycle and population dynamics in the Dunlin *Calidris alpina. Ann. Zool. Fenn.* **4:** 158–98.

Soikkeli, M. (1970). Dispersal of Dunlin *Calidris alpina* in relation to sites of birth and breeding. *Ornis Fenn.* **47:** 1–9.

Sokal, R. R. & Rohlf, F. J. (1969). *Biometry.* Freeman & Co, San Francisco.

Swingland, I. R. (1977). The social and spatial organisation of winter communal roosting in rooks (*Corvus frugilegus*). *J. Zool., Lond.* **182:** 509–28.

Ulfstrand, S. (1979). Age and plumage associated differences of behaviour among black-headed gulls *Larus ridibundus:* foraging success, conflict victoriousness and reaction to disturbance. *Oikos* **33:** 160–6.

Zwarts, L. (1974). *Vogels van het brakke getijgebied.* Bondsuitgeverij, Amsterdam. 212 pp.

Zwarts, L. (1976). Density related processes in feeding dispersion and feeding activity of teal (*Anas crecca*). *Ardea* **64:** 192–209.

Zwarts, L. & Drent, R. H. (1981). Prey depletion and the regulation of predator density: oystercatchers (*Haematopus ostralegus*) feeding on mussels (*Mytilus edulis*). In *Feeding and survival strategies of estuarine organisms*, ed. N. V. Jones & W. J. Wolff, pp. 193–216. Plenum Press, New York.

10

Differences in quality of roosting flocks of Oystercatchers

C. SWENNEN

Introduction

Most species of marine shorebirds forage in the intertidal zone during low tide and rest in monospecific groups during high tide. The same roosting places are normally used throughout the whole year, or at least for a large part of it. Birds usually forage on the flats in the immediate neighbourhood of their roost. The space available at a roost rarely, if ever, seems to be a factor limiting the number of birds using it. Yet when the tidal area is extensive, several roosts may normally be used at the same time. It is therefore usually assumed that the size of a roost depends on the size and quality of the foraging area nearby and therefore simply reflects the numbers of birds feeding in the vicinity.

Lack (1954) argued that the dispersion of roosts is primarily due to the avoidance of occupied or crowded areas by potential settlers. The behaviour of birds already present would be of no importance, the large numbers of birds themselves being the main deterrent to further settling. On the other hand, Wynne-Edwards (1962) emphasized the importance of a social hierarchy in dispersion processes. Birds would know their status relative to others and only individuals with a sufficiently high status would be able to use resources without dispute.

In extensive regions of tidal flats, considerable variation occurs in the size of area, type of substrate, duration of emergence during low tide, biomass of particular invertebrates species and total biomass of macrobenthic animals. Therefore, the quality of the foraging area adjacent to roosts is likely to vary considerably. According to Lack (1954), birds in these circumstances should occupy the optimal places first and individuals arriving later should go to suboptimal places without contest. If so, the choice between settlement in an optimal but densely populated area and a suboptimal but sparsely populated area should lead to the average bird at all roosts having similar quality and therefore the same survival rate. But if, as Wynne-Edwards (1962) argues, social rank exerts a strong influence on dispersion, birds will be distributed between roosts according to their status. The optimal roosts and foraging areas would be occupied by higher status birds, while the suboptimal places would be occupied mainly by lower status birds. The survival rates would also be expected to be different, being highest amongst birds using the best roost. Apart from the relevance of this issue to debates on population regulation and natural as against group selection, how birds of different status or quality are distributed over a spatially variable habitat has direct practical consequences for research and the conservation of shorebirds.

This chapter describes the distribution of birds of different quality over several roosts in the same area. The Oystercatcher, which is abundant in winter in the Dutch Wadden Sea, was chosen as a convenient subject for this research.

Study area and methods

The work was done on the West-Frisian islands of Texel, which has 21 km of coastline along the Dutch Wadden Sea, and Vlieland, which has 18 km of coastline. The numbers of Oystercatchers on the two islands fluctuated in parallel during the year. About 7000 birds occurred on Texel in the breeding season while 9000–18000, with a maximum of 40000, were recorded at other times (Dijksen & Dijksen 1977). About 3000 occurred on Vlieland in the breeding season, with 10000 to 25000 being found there at other times (Spaans & Swennen 1968).

Most of the island of Vlieland consists of sand dunes covered with poor vegetation. Even breeding Oystercatchers find very little food in their territories and must feed in the tidal zone at low tide (Swennen & de Bruijn 1980). A large part of Texel consists of arable fields and pastures, which can be used by Oystercatchers for foraging, but inland feeding is rare and occurs only very locally in winter.

Several Oystercatcher roosts were found on each island. Birds at the roost were counted several times during a number of winters. The dispersal of the birds over the intertidal zone was assessed from the direction of flights to and from the roosts. Samples taken from the intertidal flats allowed potential prey animals to be identified.

Samples of birds were obtained at the roosts by cannon-nets. Up to four nets measuring 15×10 m were placed so that birds could be caught from the centre and from the edges of the flock to ensure that sampling was not restricted to particular parts of the roost. The birds were weighed to the nearest gram with a large spring balance in a shelter. Weighing began one hour after the time of high water so that the birds would have little food remaining in the gut after at least three hours fasting. The birds were aged from plumage and the colour of legs, bill and eyes. Birds with buff-edged feathers, greyish legs, brown eyes, a large dark bill tip and with a somewhat worn tail and flight feathers were classified as yearlings. Birds with pink feet, a deep red iris, a bright orange eye-ring, orange-red bill without a dark tip, feathers without buff margins and with fresh tail and flight feathers were classified as adults. Birds intermediate in these respects were defined as subadults. A test of our method for ageing was done on 63 birds which had been ringed as pulli in earlier years. This showed that yearlings gave no difficulties if all the characters were used. Birds classified as subadults were in most cases in their second winter, but some appeared to be in their third or even fourth year. Birds were recorded as adults from the fourth year onwards. The wing length, bill length and presence of any abnormalities on the eyes, bill or legs were always recorded. Birds were then ringed and released.

Results

The roosts
Oystercatchers wintering in the eastern half of Vlieland used five roosts simultaneously. The distances along the shore between roosts varied between 1.5 and 3 km. Birds could be caught at all these roosts. Some roosts were also found on the western half of the island but these could not be studied because the area was difficult to reach and was used as a shooting range by tanks and jets. On Texel, the Oystercatchers used at least nine roosts, three of which were selected for detailed work.

Each roost had one and sometimes two alternative sites available at a distance of up to 500 m from its usual location which birds could use when disturbed. If the alternative places were also disturbed, the Oystercatchers usually remained in the air for a long time. In the rare cases when a flock went to a neighbouring roost, the birds stayed there for only a short time.

Sightings of colour-marked and other individually recognizable birds showed that most individuals used the same roost for many weeks or months at a time, although some did move to neighbouring roosts. The Oystercatchers fed on the intertidal flats near the roost. Observations on the direction of flights to the roost on the rising tide showed that the

Table 10.1 *Characteristics of the Oystercatcher roosts studied on Vlieland and Texel*

Place[a]	Area of tidal flat (ha) within radius of 1.5 km	Range of number of Oystercatchers outside breeding season (1968–77)
Vlieland		
V1–Posthuiskwelder	306	700–12000
V2–Oude Kooi	260	260–500
V3–Lange Paal	229	250–1500
V4–Westersche Veld	127	100–500
V5–Oostersche Veld	86	40–200
Texel		
T1–Schorren	477	1000–15000
T2–Wassenaar	116	250–1500
T3–Horntje	47	50–200

[a] See Fig. 10.1.

foraging areas used by the birds from adjacent roosts overlapped at the margins only. Where the flats were very wide, the birds dispersed outwards from the roost and did not move far in either direction along the shore.

The roosts were of unequal size (Table 10.1) but their relative sizes remained the same during 10 years. The only exception was roost V2 (Fig. 10.1) which did not attract very large numbers of birds, perhaps

Fig. 10.1. Location of the Oystercatcher roosts studied on Texel and on Vlieland. The tidal flats are shaded; the coarse stippling is the flat area above mid-tidal level. Inset shows northern part of the Netherlands.

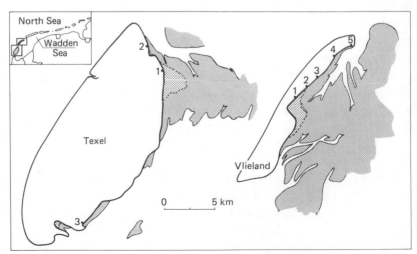

because of the semi-circle of steep dunes surrounding it. The roosts V1 and T1 had several properties in common. Both were situated on saltmarsh or former saltmarsh near the watersheds in the Dutch Wadden Sea off the islands. They were both the largest of the roosts on each island (Table 10.1). The roosts V5 and T3 were also similar to each other in being situated near the tidal inlets between the islands at places where the coast was protected by dikes. Both were the smallest of the roosts studied on each island.

There is a positive correlation between the size of the roost and the size of the adjoining intertidal flat (Table 10.1). The quality of the flats was also different. Near the largest roosts, they sloped gradually from high to low tide level. In contrast, the upper tidal zones near to the smallest roosts were truncated by a dike so that the highest parts of the flats were actually below mid-tide level. The roosts V2 to V4 and T2 were intermediate in this respect (Fig. 10.2). However, there were no differences in the potential prey species available. At all roosts the polychaetes *Arenicola marina*, *Nereis diversicolor*, *N. virens*, the molluscs *Cerastoderma edule*, *Mytilus edulis*, *Macoma balthica*, *Scrobicularia plana*, *Mya arenaria*, *Littorina littorea* and the crustacean *Carcinus maenas* were locally abundant.

The Oystercatchers

Vlieland. Samples of birds were taken at roost V1, V4 and V5 on three consecutive days. In the same year and in the same season, V2, V3 and V4 were also sampled on three consecutive days. No significant differences in age composition, weight, bill length and percentage of abnormalities were found between the replicated samples, so the data from roost V4 were combined and the series V1–V5 considered to have been taken simultaneously.

The age composition of the samples taken at the five roosts (Fig. 10.3) varied significantly ($X^2 = 169.7$, d.f. 8, $P < 0.005$). The percentages of adults were highest in V1 and decreased towards V5. The percentage of yearlings showed the opposite trend. The highest percentage of sub-adults was found at V4.

The mean weight of the birds decreased significantly ($F = 33.4$, $P < 0.005$) from roost V1 towards V5 (Fig. 10.3). Though adults are heavier than juveniles (Glutz von Blotzheim, Bauer & Bezzel 1975), the difference in age composition between roosts was not the only cause of the differences in weights because comparisons within the three age groups showed the same trend (Table 10.2).

The mean bill length of birds older than one year showed the same

Table 10.2. *Mean weights of Oystercatchers in relation to age on five roosts on Vlieland (in cases of N < 10 means are omitted)*

	V1	V2	V3	V4	V5
Yearlings	—	491	483	478	471
Subadults	—	511	517	507	—
Adults	560	519	528	532	—

Fig. 10.2. Profiles of the top 2 km of the shore adjacent to the roosts on (*a*) Texel and (*b*) Vlieland. The y-axis is the mean number of hours per tidal cycle for which a site is exposed.

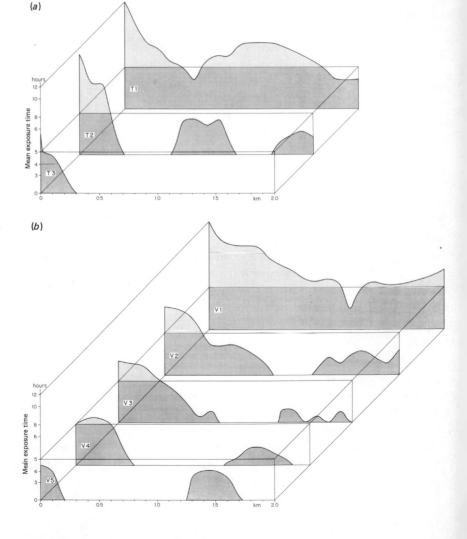

trend as the weights did, being large in V1 and small in V5 (Table 10.3). The differences between V1 and all other roosts, between V2 and V5, and between V3 and V5 were all significant (Student's t, $P < 0.01$). As in most wader species, the bills of females are larger than those of males. For the Dutch Wadden Sea, the mean bill lengths of birds older than yearlings are 71.3 mm for males and 78.4 mm for females (Glutz von Blotzheim *et al.* 1975). However, there is usually an overlap in size so the sex of individual birds cannot be determined with much certainty. But from the known mean values of the bill lengths of males and females from the western Dutch Wadden Sea, the sex ratio at the roosts could be calculated from the mean bill length of the birds caught there (Table 10.3). It is clear from this that the sexes, as well as the age groups, were distributed unevenly between the roosts.

Fig. 10.3. The age composition (*a*), mean weight (*b*) and proportion showing external abnormalities (*c*) of birds roosting at the five roosts on Vlieland. N shows the sample size.

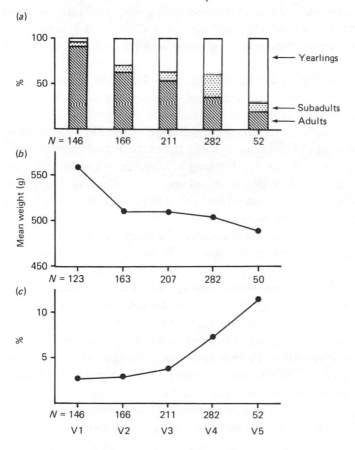

Table 10.3. *Average length of the bills of Oystercatchers over one year old at different roosts on Vlieland*

Roost	Mean length of bill	S.D.	Sample number	Calculated percentage of females
V1	76.3	5.8	121	70
V2	73.5	4.8	116	31
V3	74.3	5.8	129	42
V4	72.7	4.7	163	20
V5	70.2	5.2	14	0

Severe anatomical abnormalities were found in a significant number of Oystercatchers. For instance, some birds had one or more toes missing or carried an external tumour. Most abnormalities concerned the bill: differences in the lengths of the two mandibles exceeding 2 mm, curved mandibles or mandibles crossing each other were frequently recorded. The percentage of birds showing one or more of these abnormalities was lowest at V1 and highest at V5 (Fig. 10.3). There was no relation between the presence of these abnormalities and the age of the birds.

Texel. The results of the catches made on Texel are summarized in Table 10.4. The highest percentage of adults, the greatest mean bill length of birds over one year old and the lowest percentage of birds with external abnormalities were found at the largest roost (T1). In contrast, the smallest roost (T3) had the highest percentage of yearlings and birds with abnormalities. The mean bill length, however, was relatively high but this was at least partly due to the number of birds having overgrown, malformed bills. For parameters other than bill length, the birds caught at roost T2 were intermediate.

The weights of birds on Texel could not be compared as the weight of Oystercatchers fluctuates in the course of the year (Glutz von Blotzheim *et al.* 1975; Dare 1977) and the samples from Texel were obtained in different months.

As on Vlieland, the largest roost with the most extensive intertidal area and nearest the watershed in the Dutch Wadden Sea contained the highest proportion of adults and females and the lowest percentage of birds with abnormalities. The smallest roost with the narrow intertidal zone on both islands had the lowest percentage of adults and the highest percentage of males and birds with defects.

Mortality. The recovery rates for ringed birds were so low that they could not be used to test for possible differences in mortality between

Table 10.4. *Differences of characteristics in three roosting populations of Oystercatchers on Texel*[a]

Roost[c]		Percentage with large external abnormalities	Percentage frequencies of age classes			Mean length of bill of adults and subadults		
			Adult	Subadult	Yearling	$\bar{X}$	s.d.	(N)
T1	485	0.6	94.8[b]		5.2	75.5	5.5	90
T2	44	6.8	75.0	6.8	18.2	71.5	4.6	35
T3	63	14.3	49.2	17.4	33.3	73.5	6.3	38

[a] Catches outside the breeding season but not in the same month and not in the same year.
[b] Only a few subadults were distinguished.
[c] For size of roost see Table 10.1.

different roosts: the recovery rate of Oystercatchers ringed as full-grown birds is only 5.3 % over their whole life (Speek 1973). Furthermore, many recoveries came from outside the wintering area. Moreover, the status of individuals may have changed through time as juveniles become adults and as mandibles were broken and then regrew. As their quality changed through life, so may have their roosts. Indeed, among the 62 ringed birds recaptured in later years, about 30 % had moved to another roost within the Texel–Vlieland area.

Winter mortality is not usually very conspicuous, but dead or dying Oystercatchers were found about as frequently on the small roosts as on the large ones. This indicates a higher mortality on the smaller roosts. Confirmation came during a spell of severe frost during the last days of January and the first week of February 1976. The mean air temperature was below freezing point for 11 days. No snow fell but there was a strong easterly wind for most of the time. After one day, the higher parts of the flats were covered with ice, and the Oystercatchers did not feed there but stayed in the high-tide roost the whole day. Most birds sat close together or in sheltered spots behind vegetation. After a few days, numbers of dead Oystercatchers were noticed on the roosts. In contrast to the bodies of Redshanks, Curlews, Bar-tailed Godwits and some duck species, the dead Oystercatchers were not damaged by crows or other scavengers. All the dead Oystercatchers on the three roosts on Texel (T1–T3) were collected five or six days after the beginning of the frost and again six days later, immediately after the thaw had set in.

The number of dead Oystercatchers collected on these roosts was 380. As in the cannon-net samples, the highest percentage of adults was

Table 10.5. *Differences in mortality of Oystercatchers on three roosts on Texel during a spell of frost, February 1976*

Roost	Size of roost in the beginning and at the end of the frost period		Number found dead	Mortality (%)
T1	About 10 000	About 10 000	192	2
T2	About 1000	About 900	126	13
T3	146	80	62	42

found at T1, the highest percentage of subadults at T2, and the highest percentage of yearlings at T3. Overall, the mortality among Oyster-catchers on Texel was about 2–3 % during the two weeks of low temperatures, and similar figures were found on Vlieland (G. C. Boere, unpublished) and along the mainland coast of the Dutch Wadden Sea (J. B. Hulscher, unpublished). However, the various roosts showed noticeable differences in mortality, which was relatively low on the large roost, T1, but high on the smallest roost, T3 (Table 10.5). The dead birds tended to have low weights (36–39 % below the normal weights of the three age groups) and a high occurrence (26.6 %) of anatomical abnormalities. A high proportion (18.9 %) of the birds older than one year also had not finished moulting their primaries. Moreover, for the dead birds collected during the last week of the frost period the ratio of immatures and males had increased compared with the samples collected in the first week (Swennen & Duiven 1983).

Discussion

Oystercatchers on a roost comprised a group of birds which used the same foraging area and the same roost site for many weeks or months in winter. Although individuals did move to other roosts regularly, the majority was faithful to one roost. Furthermore, the number of ringed birds found on the same roost in later winters was much higher than the number found at a new roost. Fidelity to roost sites was therefore high.

On the West-Frisian islands, several roosts were situated along the shoreline separated by distances of 1.5–3 km. The foraging areas used by birds from neighbouring roosts only overlapped at the margins, as far as could be told from the shore. Though the birds dispersed only to a distance of between 0.5–2 km on each side of the roost, they flew up to 13 km outwards away from the shore. At such great distances, it was not possible to establish the roost from which all the birds originated.

Although the relation between the size of a roost and the size of the foraging area associated with it could not be established precisely, there was a positive correlation between roost size and the approximate size of the area used for feeding.

There is no doubt, however, that the Oystercatchers at each roost were unequal in quality. The mean bill length, wing length and weight were higher on the largest roosts than on the others. The reverse was true for the smallest roosts, and the intermediate roosts were intermediate in both respects. Only in part was this due to a separation of the sexes and age classes between roosts, and the same trends were recorded on two different islands.

The large roosts were also characterized by a high percentage of adults, a high percentage of females and a low percentage of birds with anatomical defects. In contrast, the smallest roosts had a high percentage of immatures, a high percentage of males and a high percentage of birds with abnormalities. Conceivably, some habitats may have been better for the larger adults and females while others may have been more suitable for the smaller immatures and males so that the distribution of birds was achieved solely by choice. However, it is difficult to imagine how some areas could have been especially favourable for birds with malformed bills or amputated toes. Furthermore, it is very unlikely that places with a small tidal area were better for immatures because on Vlieland the age distribution over feeding areas was exactly the opposite in the breeding season to that seen in winter (Swennen & de Bruijn 1980). This does suggest that an important proportion of the immatures did not actually have a free choice of where to settle, but went to places from which the adults did not exclude them.

The different mortality rates do suggest that there must have been crucial differences in the quality of either the roosting sites or of the nearby foraging areas, or both. Probably, the largest foraging areas were also the best in winter. Besides the difference in size of intertidal area, the most striking difference between feeding areas on both islands was the difference in level. The tidal flats near the large roosts descended gradually from high to low tide level whereas those near the smallest roosts only extended from about mid-tide to low tide level. The other roosts were again intermediate.

All the important prey species were found in each of the foraging areas. The highest biomass of benthic invertebrates is found around and somewhat below mid-tide level (Beukema 1976), so all the foraging areas possessed a rich zone where the Oystercatchers could forage for about five hours during low tide. Though organisms higher up the beach were available for much longer periods, they were, on average, smaller. Accordingly, the high level areas near to the large roosts were intensively used by Oystercatchers only when foraging at a lower level of the beach was impossible. This happened regularly when the sea level was raised by strong westerly winds. On these occasions, the Oystercatchers on the smaller roosts remained the whole day at their roost while birds from the larger roosts were able to forage for at least a few hours. The frequent raising of the sea level in the winter of 1976 was probably crucial to the high mortality during the cold spell that followed (Swennen & Duiven 1983).

The results of this study indicate that the distribution of Oystercatchers between roosts was not established in a free way as would be supposed by Lack (1954). There were distinct differences in the quality of birds using the different roosts. In the best sites, with flats extending over all tidal levels, the average quality of the birds was high. At the less favourable sites, with a small intertidal area at only the lower tidal levels, the average quality of the birds was low. Clearly, differences in mortality of such magnitude occurring over such small distances would not have happened if all the Oystercatchers had been free to live in the best areas. The differences in spatial distribution were clearly not related to a single feature (age, sex, defects) because birds sharing particular combinations of these characters were only partially segregated. It is therefore likely that individuals with about the same social rank occurred together and that low-ranking birds were forced in some way to leave the best places, as was proposed by Wynne-Edwards (1962) for birds in general and by Goss-Custard & Durell (Chapter 11 of this volume) for Oystercatchers in particular. Low-status birds may have done better by moving to less favourable areas because the competition from high-status birds was less there.

The idea that the partial segregation of age groups and sexes in other marine birds is caused by a social hierarchy receives some support from the studies of Spaans (1970) and of Monaghan (1980) on Herring Gulls in which adults and males are usually dominant. Similarly, adult Eider Ducks are dominant to the immatures (author's personal observation).

The fact that there were considerable differences between the characteristics of birds from different roosts has practical consequences for research. It means that bias is likely to arise when attempts are made to measure age composition, sex ratio, mean weight, mean wing length, mean bill length, mean condition, mean food consumption, survival, etc. in a wintering population and to compare the results with data from elsewhere. It is certainly not sufficient to collect most data from the smaller, more easily studied roosts because the birds there may not be typical.

References

Beukema, J. J. (1976). Biomass and species richness of the macrobenthic animals living on the tidal flats of the Dutch Waddenzee. *Neth. J. Sea Res.* **10**(2): 236–61.

Dare, P. J. (1977). Seasonal changes in body-weight of oystercatchers *Haematopus ostralegus*. *Ibis* **119**: 494–506.

Dijksen, A. J. & Dijksen, L. J. (1977). *Texel vogeleiland*. Thieme, Zutphen.

Glutz von Blotzheim, U. N., Bauer, K. M. & Bezzel, E. (1975). *Handbuch der Vögel Mitteleuropas*, vol. 6. Akademische Verlagsgesellschaft, Wiesbaden.

Heppleston, P. B. & Kerridge, D. F. (1970). Sexing oystercatchers from bill measurements. *Bird Study* **17**: 48–9.

Lack, D. (1954). *The natural regulation of animal numbers*. Clarendon Press, Oxford.

Monaghan, P. (1980). Dominance and dispersal between feeding sites in the Herring Gull (*Larus argentatus*). *Anim. Behav.* **28**: 521–7.

Spaans, A. L. (1970). On the feeding ecology of the Herring Gull *Larus argentatus* Pont. in the northern part of the Netherlands. *Ardea* **59**: 75–188.

Spaans, A. L. & Swennen, C. (1968). De vogels van Vlieland. *Wetenschappelijke Mededelingen Koninklijke Nederlandse Natuurhistorische Vereniging* **75**: 1 104.

Speek, B. J. (1973). Ringverslag van het Vogeltrekstation. *Limosa* **46**: 136–65.

Swennen, C. & de Bruijn, L. L. M. (1980). De dichtheid van broedterritoria van de Scholekster *Haematopus ostralegus* op Vlieland. *Limosa* **53**: 85–90.

Swennen, C. & Duiven, P. (1983). Characteristics of Oystercatchers killed by cold-stress in the Dutch Wadden Sea area. *Ardea* **71**: 155–9.

Wynne-Edwards, V. C. (1962). *Animal dispersion in relation to social behaviour*. Oliver & Boyd, Edinburgh and London.

11

Feeding ecology, winter mortality and the population dynamics of Oystercatchers on the Exe estuary

J. D. GOSS-CUSTARD &
S. E. A. le V. dit DURELL

Introduction

Two questions are frequently asked about wader populations wintering on estuaries in northwest Europe. One is whether a loss of intertidal feeding area from various causes is likely to reduce the numbers of birds, either in the local population through increased emigration or in the entire population through greater mortality and/or reduced breeding success. Less frequently, the issue is whether a wader, usually the Oystercatcher *Haematopus ostralegus*, is a pest of commercial shell-fisheries and, if so, whether practical measures can be devised to reduce their numbers. To answer both questions requires an understanding of the way in which the populations interact with each other. Conservationists need to know whether reductions in food will reduce bird numbers, and fishermen need to know whether reductions in bird numbers will improve the fishery.

A great deal of work has been done in the last 25 years, with most research recently being focussed on conservation. Despite this, confident predictions are still difficult to provide. This is because of the complexity of population processes, the numbers of species involved, their differing life styles and, in the case of waders, their migratory habits. Waders that breed in one place may spread over a huge area in winter while birds that winter on one estuary may breed in several countries. It is difficult to study a group of birds which can be identified as a discrete population, and this has hindered progress. Furthermore, much of the research has been done for only the few years of a feasibility study on huge intertidal areas threatened by commercial developments. The study areas have often been too large and the study periods too short for the detailed work required. Nor is it safe to assume that the population size in winter is determined by the resources available at that time because numbers may be influenced more by factors operating on the breeding areas, or even along the migration routes. Whereas the loss of preferred habitats in resident bird species is likely to reduce the population because it will normally have reached the level set by the resources available, this may not be assumed for wintering populations of waders.

Our long-term study on Oystercatchers and their main food, *Mytilus edulis*, began in 1976 on the Exe estuary in southwest England. The bird population is small enough to allow individuals to be followed and both study area and prey population are of manageable size (Goss-Custard *et al*. 1980). Though the overall aim is to investigate the role which Oystercatchers and mussels play in each other's population dynamics, this paper discusses the possible effect of only winter mortality and food supply on bird numbers. The main findings are described briefly, and the details of methods and results are given in Goss-Custard (1981*a*), Goss-Custard & Durell (1983) and Goss-Custard *et al*. (1980, 1981, 1983*a*,*b*,*c*).

The Oystercatcher population

The Exe estuary is approximately 10 km long and mostly 2 km wide (Fig. 11.1). The 31 mussel beds occur mostly in the lower reaches. Some Oystercatchers feed on the nearby coast and fields but are included in the study population as most of their activities centre on the estuary. Very few birds move between the River Exe and the River Teign, about 8 km to the west. Most individuals remain on the Exe for long periods and 80–90%, depending on age, return there each winter. Within one winter, the turnover is low and there is an easily definable local population of birds.

The numbers of Oystercatchers in the study area vary between seasons but not much between years (Fig. 11.2). There are only a few hundred, mainly immature birds (aged from one to approximately four years old) on the Exe in spring and summer when the adults are away breeding in the Netherlands, Scandinavia and northern Britain. The adults return between mid-July and mid-September, with the juveniles, the young birds produced that year (though not necessarily by adults from the Exe), arriving somewhat later. The population peaks in late September and may be followed by a small decline during October. Total numbers change little until adults start to migrate in early February. Most birds have left by mid-April.

The change in numbers between autumn and spring is small because few birds leave and mortality is low. The numbers dying in autumn and winter can be calculated because the proportion of the dead ringed birds reported by the general public has been estimated. Most of the birds that disappear then are known to have died, although some immatures do move to other estuaries. Mortality on the Exe in autumn and winter declines sharply with age (Fig. 11.3). In contrast, the proportion of birds known to die in spring and summer elsewhere increases with age. The

Fig. 11.1. The Exe estuary in southwest England, and the numbered mussel beds.

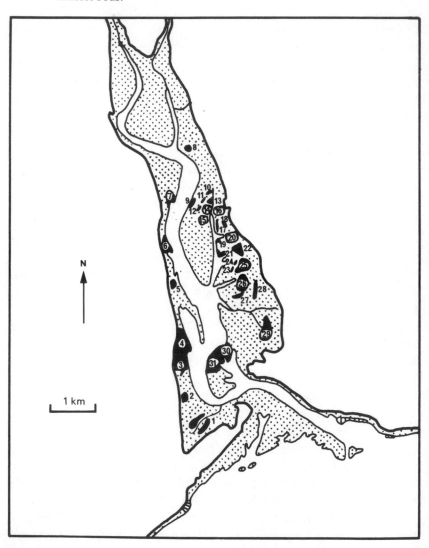

estimates shown in Fig. 11.3 are probably rather low because of the difficulty in calculating the proportion of ringed corpses reported at this time of year. The annual mortality for older birds is estimated to be 4–5%, which is a little low compared with most of the other estimates (2.5–14%) based on return rates to breeding areas (e.g. Harris 1975). But the main trends in mortality are clear. It is highest in young birds, and mostly occurs on the Exe in autumn and winter. Annual mortality declines as the birds mature, and an increasing proportion of it occurs elsewhere. Whereas 12% of juveniles in the study area die in autumn and winter, less than 2% of the adults do.

Fig. 11.2. Mean numbers (± s.e.) of Oystercatchers on the study area each month: data for all age groups (solid squares) from July 1976 to June 1981; data for adults (solid circles), immatures (open circles) and juveniles (triangles) from July 1976 to November 1979.

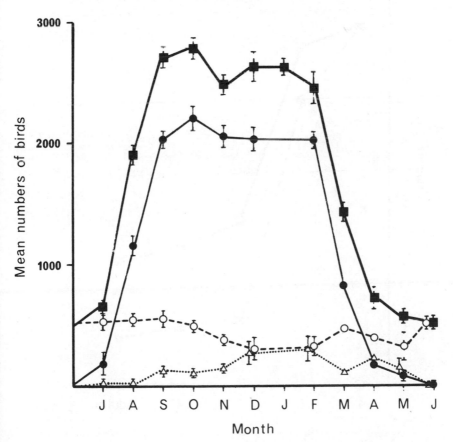

Feeding ecology

Most Oystercatchers roost at high water along the coast, mainly on Dawlish Warren (Fig. 11.4), though in winter about 25% feed in fields around the estuary. At low tide, most birds feed on the mussel beds (Fig. 11.5). The remainder forage on the flats within the estuary and, particularly in winter when disturbance is least, along the coast. Very few feed in the fields.

Most prey eaten on the mussel beds are mussels, together with other bivalves, mainly *Cerastoderma* and *Spisula* spp., periwinkles (*Littorina*

Fig. 11.3. The minimum proportion of birds of each age dying on the Exe in autumn and winter (solid squares), elsewhere in spring and summer (open squares), and in both periods combined (half-solid squares).

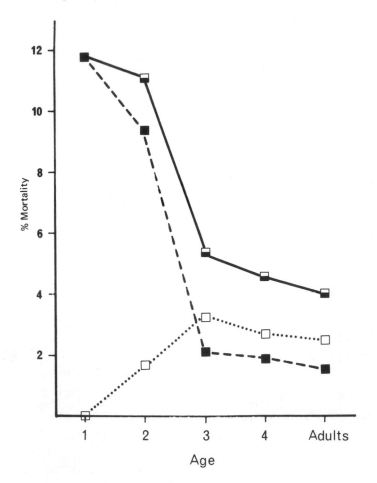

spp.) and crabs (*Carcinus maenas*). On the mudflats, Oystercatchers mainly eat the bivalve *Scrobicularia plana* in winter and the polychaete worm *Nereis diversicolor* at other times of year. Cockles are eaten on the sandflats, *Spisula* and *Mytilus* along the coast and earthworms (Lumbricidae) in the fields.

The diet changes with age. At low water, many juveniles feed on *Scrobicularia plana* and *N. diversicolor* from the mudflats, *Spisula* from the beaches and, particularly at high tide, earthworms from the fields. Adults mainly eat mussels and seldom occur in the fields. From being a minority diet in young birds, mussels become the most important food in adults (Fig. 11.6).

Studies of colour-marked birds have shown that most individuals have a restricted diet compared with the range eaten by the population as a whole. Most adults and older immatures eat mainly mussels and very few other prey. But a small minority of adults specialize in *Littorina* at low water and may feed in the fields at high tide. Others eat only *N. diversicolor* in autumn and spring and *Scrobicularia plana* and earthworms in the winter. Some eat only *Cerastoderma*. These differences

Fig. 11.4. Seasonal changes in the proportion of the population at the roosts or feeding in the fields. Data for 1976–80.

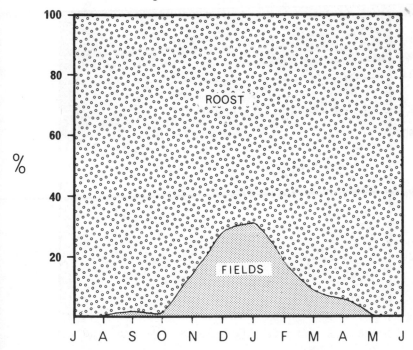

persist for long periods, even from winter to winter. Similar specialization occurs amongst younger birds. But considering the population as a whole, most birds obtain the major proportion of their food from mussels because most of the population are adults and most adults specialize on mussels.

Feeding patterns on the mussel beds

Preference for different beds

The birds are not distributed evenly over the 31 mussel beds on the Exe estuary. In spring and early summer over half of the young birds then present feeds on beds 30 and 31 which are situated near the mouth of the river (Fig. 11.7). Many of the other mussel beds have no birds on them at all. But as the population feeding on mussels increases sixfold during the late summer and autumn as the adults return, the birds spread out over the mussel beds until almost all are occupied. The number of birds feeding on beds 30 and 31 increases but not in

Fig. 11.5. The proportion of birds feeding on each habitat at low tide throughout the year. Data for 1976–80.

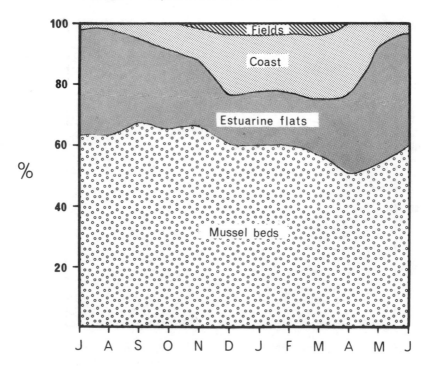

proportion to the rise in the population as a whole. Consequently, the proportion of the Oystercatcher population feeding there declines from over 50 to almost 20%.

In general, the beds, like 30 and 31, which are occupied early during the build-up also have relatively high densities throughout autumn and winter. Beds can therefore be ranked according to the birds' preference for them on the basis of these two criteria. This enabled the mean rank of the beds used by the birds to be calculated, and this changed throughout the year in line with the changes in bird numbers (Fig. 11.8). The average Oystercatcher feeds on high-ranked beds in summer when the population is low, but on less preferred beds in autumn and winter when the population is high.

Studies on several wader species (Goss-Custard 1970, 1977a,b; Smith 1975; Zwarts & Drent 1981; Sutherland 1983) suggest that the birds prefer to feed where their intake rate is highest. While there is some evidence for this from Oystercatchers on the Exe, net intake rate may be the important factor. Food is not unusually abundant on beds 30 and 31, yet these are the most preferred beds. The features which make them preferred are probably their proximity to the main roost and the firmness of the substrate, neither of which seems to affect intake rate but which could significantly influence the amount of energy expended in

Fig. 11.6. The proportion of the sightings of individually marked birds of known age on mussel beds (solid circles) and the proportion of records where the diet was determined where the bird was eating mussels (open circles). A = autumn, W = winter, S = spring.

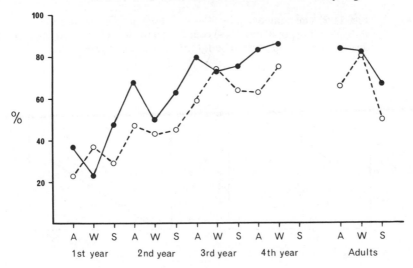

foraging. This may have a critical effect on the net rate of energy intake and hence the attraction of a feeding area. We have not tested this but the preferred beds probably do provide the best feeding conditions.

Competition for mussel beds

If some beds are preferred, why do not all the birds feed there? The birds must compete for the best feeding areas, with competition intensifying as the population builds up so that an increasing proportion is forced to feed in less preferred areas. To see how this might work, the birds' social behaviour on mussel beds was studied.

Most work was done during two winters on bed 4, one of the highest ranked beds. Sixty-one colour-ringed individuals were seen during the first winter, but just over half on only one or two occasions. These birds mostly fed elsewhere, and were classed as transients. The remaining 27 birds, seen on between 5 and all the 21 days that the bed was watched, were called 'regular visitors'. They fed in very small areas of the mussel bed whenever they were present, and returned to the same area for the second winter. They did not defend exclusive feeding territories, and feeding ranges overlapped considerably. Aggressive encounters were, however, very frequent, especially when bird density was high on receding tides. Approximately 60 % of the encounters involved attempts to steal mussels and many of the other encounters may have involved feeding sites rather than particular mussels. As Vines (1980) showed for the Ythan, most encounters were over food.

There were considerable individual differences in aggressiveness. The

Fig. 11.7. The numbers of birds (—) and the proportion of the mussel-eating population (---) counted on the most preferred beds (30 and 31) and least preferred beds (1, 13, 14, 22) on the estuary.

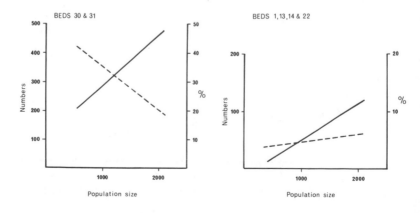

most aggressive birds attacked another once every 5 min at low tide. They were relatively infrequently attacked themselves, and won most encounters whether attacking or defending a mussel or feeding site (Fig. 11.9). At the other extreme, some individuals seldom attacked others but were themselves frequently attacked and usually lost. There was a linear dominance hierarchy amongst any group of birds whose feeding ranges overlapped (Ens & Goss-Custard 1984). Most birds both stole mussels and had mussels stolen from them, but the net gain or loss of mussels in such encounters varied considerably. Aggressive birds stole far more than they lost, and some obtained up to 20 % of their food this way. In contrast, sub-dominant individuals suffered a net loss in encounters over mussels. Since stealing mussels was profitable, the dominants significantly increased their overall rate of intake by stealing, though the effect was not very large.

Dominant birds gained most when bird density was highest, usually when the area of bed available for feeding was least. The intake rate of the most dominant birds was little affected at high densities, whereas that of most of the sub-dominants seemed to be considerably depressed (Ens & Goss-Custard 1984). Interestingly, rather little of the reduced intake rate was due to more mussels being stolen from them: they simply found fewer. They may have been distracted from looking for

Fig. 11.8. The mean rank of the mussel beds used by the population in spring and summer, autumn and winter. A high score indicates that most birds were on the most preferred beds (high score = 6, low score = 4).

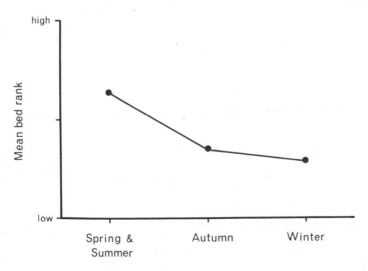

mussels because of the presence of dominants nearby. Alternatively, in attempts to avoid dominant birds (Vines 1980), the sub-dominants may have been forced to feed in the less profitable parts of their feeding range. But whatever the mechanism, this work suggested that the effect of high bird densities on intake rate did vary between individuals according to their status.

These results suggest how we may test the idea that competition for the preferred beds explains the sequence in which they are occupied by the returning birds in late summer. If competition does occur for the most preferred beds, the sub-dominant young birds feeding on beds 30 and 31 in the summer would be expected to move away in the autumn on the adults' return, and this does occur (Fig. 11. 10*a*). About 20% of these young birds then feed on the mud, but most move to less preferred mussel beds. Consequently, in winter the proportion of young birds on the low-ranked beds is much higher than on the preferred beds (Fig. 11.10*b*). The young birds return gradually to the preferred beds in spring and summer, only to be replaced again by the adults in autumn. However, an increasing proportion of birds remain on the preferred beds in winter as they grow older so that the average rank of the beds used by birds increases with their age.

Fig. 11.9. The relationship between the aggressiveness of a bird and the proportion of encounters it wins.

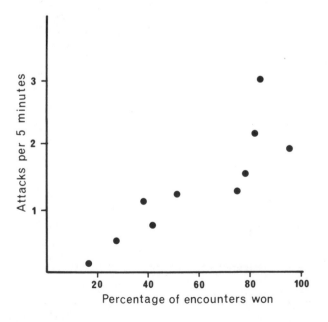

Another test of the competition hypothesis is that the amount of feeding space available on the mussel beds should become increasingly important in determining bird density as the population increases. As Fig. 11.11 shows, this is the case. The amount of feeding space on a mussel bed is described by the abundance of the mussels over 30 mm long from which Oystercatchers obtain most of their food. At low population levels, bird density is only correlated with a bed's distance from the roost, hardness of its substrate and the thickness of the mussel shells: the density and size of the mussels over 30 mm long are unimportant. But as numbers build up, bird density becomes increasingly closely correlated with both these variables. The need to find feeding space plays a bigger role in the distribution of the Oystercatchers as numbers increase.

Mortality of young birds

These results raise the possibility that the relatively heavy mortality of first- and second-year birds in autumn and early winter is partly a result of competition with adults for the best feeding areas. Unfortunately, it is difficult to test this idea directly. There are only a few hundred first- and second-winter birds on the Exe, and only 10%–15% of them die. A very small proportion is found in a condition which allows a post-mortem to be done. However, young Oystercatchers are lighter than older birds throughout autumn and winter on the Exe as elsewhere (Dare 1977), which may indicate that their intake rate

Fig. 11.10. (*a*) The numbers of adult and immature birds on beds 30 and 31 during late summer and autumn and (*b*) the proportion of immatures amongst birds on beds of different rank. A high score indicates a highly preferred bed.

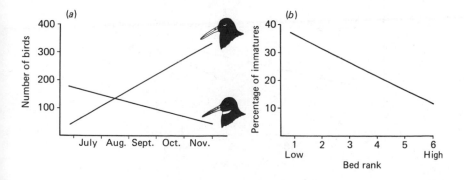

is unusually low. Norton-Griffiths (1968) showed that young Oyster-catchers are less efficient than adults, but this possibility has not yet been tested on the Exe.

If young birds do feed less well than older ones, it is important to determine how much this is due to inexperience and how much to competition with dominant adults, as both could be involved simul-taneously. If competition does contribute to the mortality of young birds, any loss of feeding areas could increase the mortality amongst them. What effect might this have on the size of the Oystercatcher population as a whole? This question is explored using a simple mathematical model of the population of Oystercatchers wintering on the Exe.

The main features of the model, which is described in detail in Goss-Custard (1980, 1981*b*), are as follows. It is assumed that produc-tion is density-dependent because territoriality limits the density of breeding birds. This is based largely on the work of Harris (1970, 1975) on an island population of Oystercatchers, and further tests on mainland birds are desirable. An increasing proportion of the birds fail to obtain

Fig. 11.11. The partial correlation coefficients between the density of Oystercatchers and each of five attributes of the mussel beds in relation to population size.

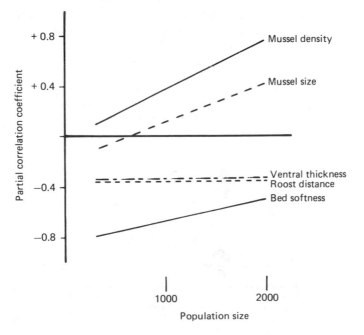

Table 11.1. *Parameter values used in the Oystercatcher population model (Goss-Custard 1980, 1981b). Where a range is given, values were varied at random within the range. Unless stated, mortality was density-independent*

Season	Age class	Process	Parameter value
Breeding	Adults	Number of breeding pairs	Intercept: -1.72; slope: 0.625
	Breeding adults	Mortality	0.025–0.075
	Non-breeding adults	Mortality	0.025–0.075
	2–4 year olds	Mortality	0.025–0.075
	Eggs	Clutch size	2.500–3.000
	Eggs	Mortality	0.350–0.850
	Chicks	Mortality	0.400–0.800
	Fledglings	Mortality	0.200–0.400
Autumn	Juveniles	Mortality	Variable – see text
	2nd year	Mortality	Variable – see text
	3rd, 4th year and adults	Mortality	0.01
Winter	Juveniles	Mortality	Variable – see text
	2nd year	Mortality	Variable – see text
	3rd, 4th year and adults	Mortality	0.015

territories as the population increases, but the contraction of territory size as competition increases is incorporated in the model as this is how territory size may vary in nature (Vines 1979). The number of territories depends on the density at which potential breeders begin to be excluded (i.e. the intercept in the equation given in Goss-Custard (1980)) and the compressibility of territories (i.e. the slope). In this exercise, intermediate values of slope and intercept were used which generated a stable population size comparable to that occurring on the Exe. The other parameter values are listed in Table 11.1 and were based on a literature review (Goss-Custard 1981b). Important features are that mortality from the egg stage onwards was variable, but usually density-independent, and from the fledgling stage onwards could occur in autumn, winter, spring or summer. Each year class was treated separately until the fifth year when all birds were assumed to be adult and able to breed. Breeding pairs that died were replaced by other birds, which bred. As has been shown for the Exe population, most of the annual mortality occurred away from the Exe, except in the case of the first- and second-year birds.

The way density-dependent mortality in young birds was varied in the model is illustrated in Fig. 11.12. It is assumed that the mortality rate of the young birds depends on the size of the whole population but that it only becomes density-dependent above a certain population size when competition begins to take effect. Below this level, the mortality rate was density-independent at a value of 10%. The effect of increasing competition through removing food supplies may be modelled by increasing the slope in the density-dependent region of the curve or by moving it to the left so that it begins to take effect at lower population sizes. In practice, both may happen simultaneously. In these simulations, however, the slope was held constant and density-dependence started to operate at progressively lower population sizes. The slope chosen was a moderate value and other simulations showed that the main conclusion of this work was not much affected by varying it within its probable range. Similarly, setting the mortality of the more experienced second-year birds to be half that of the first-year birds did not seriously affect the conclusion. As Fig. 11.13 shows, decreasing the population size at which mortality becomes density-dependent has a substantial effect on the overall size of the population, even though young birds make up only a small proportion of the whole population and mortality never rises to very high levels.

Discussion

This study suggests that increasing the mortality of a small and vulnerable minority of the population in winter may have a substantial

Fig. 11.12. The way in which density-dependent mortality amongst the young birds (first and second year) is expressed in the model. Total population size refers to all age classes combined.

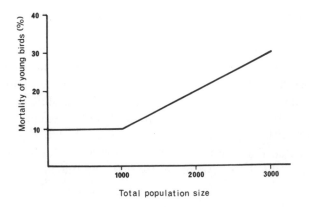

effect on stable population size. This is true even though most of the mortality in the population as a whole occurs away from the Exe and in other age groups, particularly in eggs, hatchlings and fledglings, and the main source of regulation is provided by competition for breeding territories. By reducing the numbers of recruits to the future adult population, increasing the winter mortality of young birds can reduce population size even though subsequent competition for territories on the breeding grounds will also be reduced. This suggests that it would be worthwhile testing the assumptions of the model in the field. The most important are (i) that being sub-dominant to older birds, rather than simply being inexperienced, actually contributes to any difficulty the young ones have in obtaining food (see van der Have *et al.*, Chapter 9 of this volume), including prey species other than mussels which have not yet been studied, and (ii) that food shortage is at least a contributory factor to the autumn and winter mortality of young birds (see Swennen, Chapter 10 of this volume). Unless the latter is true, the penalty for being sub-dominant may be trivial and incur only insignificant risks.

The advantages of using a mathematical model to explore the effects of increased mortality have been discussed elsewhere (Goss-Custard 1980), but they may be mentioned briefly. The general point is that a

Fig. 11.13. The effect on stable population size of moving the density-dependent region of the curve in Fig. 11.12 to the left while holding its slope constant.

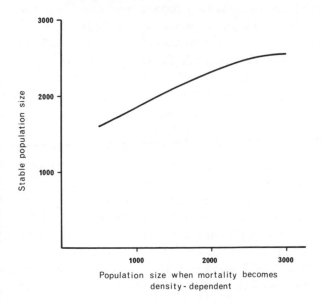

model provides a means by which it is possible to think quantitatively. Since the problem being considered concerns bird numbers, it is a numerical problem for which quantitative thinking is likely to be beneficial. It would not be easy with words alone to predict the outcome of changing the value of a quantity in a system that contains several variables. By testing the sensitivity of the conclusions to changes in the assumptions about the way the system works, the model helps to identify what may be the particularly crucial questions for further testing in the field.

A mathematical model also emphasizes the interactive and dynamic properties of populations. All the mortalities operating on the population and their degree of density-dependence and the sequence in which they act will affect the mean population size, its amplitude of fluctuations and its stability (e.g. Varley, Gradwell & Hassell 1973). Williamson (1972) shows how increasing any mortality rate in the population will reduce stable population size except, presumably, on the rare occasions when an exactly compensating density-dependent mortality operates afterwards. With such dynamic systems, it is unlikely that static concepts, such as the much used 'carrying capacity', will be adequate to cope with the complexity of nature. This concept implies that some 'limiting' factor in the habitat sets a ceiling on the numbers of birds that can be supported there. If numbers rise above this level, the surplus must emigrate or die. This may happen, but probably only under very special circumstances in winter. For instance, if waders lived in territories which were incompressible beyond a certain point and those birds that failed to obtain them died, or emigrated, and if the lower limit of territory size did not vary in response to annual variations in the food supply or any other factor, the concept might be useful. But field work shows these possibilities to be most unlikely. Where territories occur on the wintering areas, only a minority of the population occupies them (Goss-Custard 1970; Myers, Connors & Pitelka 1979*a*; Townshend *et al.*, Chapter 8 of this volume), and the occurrence of territories, and their sizes, are responsive to variations in both bird numbers and food supply (Myers *et al.* 1979*b*). In the much more numerous non-territorial systems, there is no evidence to suggest that density reaches a ceiling level in any but the most preferred areas (Goss-Custard 1977*a,b*). In the Exe, for example, the density of birds on even the most preferred beds continued to rise as the population increased. The study there confirms that it is probably more realistic to ask whether a decrease in food supplies or an increase in bird numbers is likely to increase mortality and/or to reduce later breeding success by increasing interference and prey depletion or by forcing an increasing proportion of the population to feed in the poorer parts of the food gradient (Goss-Custard & Charman 1976; Goss-Custard 1977a). Instead of asking whether the mortality of the young birds is too high to enable the population to reach a level set by the carrying capacity of the estuary, it is useful to view the rate of survival of these young birds as contributing to that level. The crucial questions are, then, whether mortality (or subsequent breeding success) is dependent or independent of bird density, by how much the removal of the feeding habitat would increase mortality and what effect this would have on stable population size in the long term.

Acknowledgements

Many people have contributed to this long-term study, some directly and some through discussion. We would particularly like to thank those who have worked with us on the Exe, our colleagues Drs S. McGrorty and C. J. Reading, and in other capacities, Sherman Boates, Bruno Ens, Justine Selman and Bill Sutherland. We are also very grateful to Dr D. Jenkins for commenting on the script.

References

Dare, P. J. (1977). Seasonal changes in body-weight of oystercatchers *Haematopus ostralegus*. *Ibis* 119: 494–500.

Ens, B. & Goss-Custard, J. D. (1984). Interference among Oystercatchers *Haematopus ostralegus* feeding on mussels *Mytilus edulis* on the Exe estuary. *J. anim. Ecol.* 53: 217–32.

Goss-Custard, J. D. (1970). The responses of redshank (*Tringa totanus* (L.)) to spatial variations in the density of their prey. *J. anim. Ecol.* 39: 91–113.

Goss-Custard, J. D. (1977a). The ecology of the Wash. III. Density-related behaviour and the possible effects of a loss of feeding grounds on wading birds (Charadrii). *J. appl. Ecol.* 14: 721–34.

Goss-Custard, J. D. (1977b). Predator responses and prey mortality in redshank, *Tringa totanus*, and a preferred prey, *Corophium volutator*. *J. anim. Ecol.* 46: 21–35.

Goss-Custard, J. D. (1980). Competition for food and interference among waders. *Ardea* 68: 31–52.

Goss-Custard, J. D. (1981a). Oystercatcher counts at roosts and at feeding grounds. *Br. Birds* 74: 197–9.

Goss-Custard, J. D. (1981b). Role of winter food supplies in the population ecology of common British wading birds. *Vert. orn. Ges. Bayern* 23: 125–46.

Goss-Custard, J. D. & Charman, K. (1976). Predicting how many wintering waterfowl an area can support. *Wildfowl* 27: 157–8.

Goss-Custard, J. D. & Durell, S. (1983). Individual and age-differences in the feeding ecology of oystercatchers, *Haematopus ostralegus*, wintering on the Exe estuary, S. Devon. *Ibis* 125: 155–71.

Goss-Custard, J. D., Durell, S., McGrorty, S., Reading, C. J. & Clarke, R. T. (1981). Factors affecting the occupation of mussel *Mytilus edulis* beds by oystercatchers *Haematopus ostralegus* on the Exe estuary, S. Devon. In *Feeding and survival strategies of estuarine organisms*, ed. N. V. Jones & W. J. Wolff, pp. 217–29. Plenum Press, New York.

Goss-Custard, J. D., Durell, S., Sitters, H. & Swinfen, R. (1983a). Age-structure and survival of a wintering population of oystercatchers, *Haematopus ostralegus*. *Bird Study* **29:** 83–98.

Goss-Custard, J. D., Durell, S., McGrorty, S. & Reading, C. J. (1983*b*). Use of mussel beds *Mytilus edulis* L. by oystercatchers *Haematopus ostralegus* L. according to age and population size. *J. anim. Ecol.* **51:** 543–54.

Goss-Custard, J. D., Durell, S. & Ens, B. J. (1983*c*). Individual differences in aggressiveness and food stealing among wintering oystercatchers, *Haematopus ostralegus* L. *Anim. Behav.* **30:** 917–28.

Goss-Custard, J. D., McGrorty, S., Reading, C. J. & Durell, S. (1980). Oystercatchers and mussels on the Exe estuary. *Devonshire Ass. Spec. Vol.* **2:** 161–85.

Harris, M. P. (1970). Territory limiting the size of the breeding population of the oystercatcher (*Haematopus ostralegus*) – a removal experiment. *J. anim. Ecol.* **39:** 707–13.

Harris, M. P. (1975). Skokholm oystercatchers and the Burry Inlet. *Rep. Skokholm Bird Observ.* 1974: 17–19.

Myers, J. P., Connors, P. G. & Pitelka, F. A. (1979*a*). Territoriality in non-breeding shorebirds. In *Shorebirds in marine environments*, ed. F. A. Pitelka, pp. 231–46. Cooper Ornithological Society.

Myers, J. P., Connors, P. G. & Pitelka, F. A. (1979*b*). Territory size in wintering sanderlings: the effects of prey abundance and intruder density. *Auk* **96:** 551–61.

Norton-Griffiths, M. (1968). The feeding behaviour of the oystercatcher *Haematopus ostralegus*. Unpublished PhD thesis, Department of Zoology, University of Oxford.

Smith, P. S. (1975). A study of the winter feeding ecology and behaviour of the bar-tailed godwit (*Limosa lapponica*). Unpublished PhD thesis, University of Durham, South Road, Durham DH1 3LE, England.

Sutherland, W. J. (1983). Spatial variations in the predation of cockles by oystercatchers at Traeth Melynog, Anglesey. II. The pattern of predation. *J. anim. Ecol.* **51:** 491–500.

Varley, G. C., Gradwell, G. R. & Hassell, M. P. (1973). *Insect population ecology.* Blackwell, Oxford.

Vines, G. (1979). Spatial distributions of territorial aggressiveness in oystercatchers, *Haematopus ostralegus* L. *Anim. Behav.* **27:** 300–8.

Vines, G. (1980). Spatial consequences of aggressive behaviour in flocks of oystercatchers, *Haematopus ostralegus* L. *Anim. Behav.* **28:** 1175–83.

Williamson, M. (1972). *The analysis of biological populations.* Edward Arnold, London.

Zwarts, L. & Drent, R. H. (1981). Prey depletion and the regulation of predator density: oystercatchers (*Haematopus ostralegus*) feeding on mussels (*Mytilus edulis*). In *Feeding and survival strategies of estuarine organisms*, ed. N. V. Jones & W. J. Wolff, pp. 193–216. Plenum Press, New York.

PART 3

The significance of specific areas on the Palaearctic–African migration routes of waders

Introduction

W. G. HALE

Over the past 20 years a great deal of work on wading birds in winter has provided information on many aspects of their biology. We now know where to expect concentrations of waders in western Europe and broadly speaking where they have come from and where they are going to. We know what food they take and roughly how much of it. We know when and where most species moult and when different populations move north and south. We know something of their build-up of reserves for these journeys and something of the seasonal mortality. All this has been possible through the cooperative efforts of many amateur ornithologists and a group of professional researchers, financed largely because of the conservation importance of the coastal areas used by waders. Because waders use these areas mainly outside the breeding season, much the greater part of the research effort has been concentrated on the non-breeding season and particularly in the area where waders occur most commonly. In this part of the book the more important areas for migrating and wintering waders are reviewed briefly and its scope extends into North Africa, to Morocco and Mauritania where there are important concentrations. Even within the period of the blooming of wader studies as a field of research, changes in our estuaries and other coastal wetlands have resulted in changes in wader behaviour. Some of these changes have been brought about by man, others by the vicissitudes of the environment; waders live in a changing world and this part is concluded by a review of general changes in wader habitat.

The purpose of the wader meetings (this volume records the proceedings of the fifth) is to bring together current workers to discuss their work and ambitions. Apart from the exchange of ideas, their most important aspect is to look to the future and to determine in what direction wader research should go. It is therefore appropriate that each chapter looks to the future but not unreasonably this tends to be within the context of the chapter content. The fact must not be lost to view that in some areas of research we may be looking in completely the wrong place for an answer. Numbers on estuaries may well reflect a population regulated by circumstances prevailing on the breeding grounds and this has been argued elsewhere (Hale 1980).

The Wadden Sea is a very important area for wading birds and this is reflected here by three chapters devoted to the area. The Dutch Wadden Sea, described by Wolff & Smit, is at present the most intensively worked part of the area and of particular interest here is the estimate of consumption of invertebrate production by demersal fish and other invertebrates. Together these groups take roughly twice as much of the biomass of production as the birds, and all predators consume between 55 and 70 % of the annual production. In contrast, waders on the Banc D'Arguin are estimated to take nearly the whole of the February standing crop during the period of their presence (Engelmoer, Piersma, Altenburg & Mes) of 40–60 days, allowing nothing for predation by other groups of organisms. In both areas waders are calculated to be closer to their maximum theoretical limits than in other areas where work has been carried out (Hale 1980).

The Danish Wadden Sea (Laursen & Frikke) is interesting because it is the most northerly area where concentrations of waders occur on the continental mainland and like the German Wadden Sea (Prokosch) experiences harder winters than the Dutch Wadden Sea. From the conservation point of view it is of great importance to be able to consider the Wadden Sea as a whole and whilst there are obvious political difficulties in implementing conservation strategies across national boundaries, the international cooperation in the basic ecological study of the area is a basis from which political cooperation may spring. In the end it is the political decisions which govern conservation strategy and ideally these are based on scientific recommendations; if we are able to make these on a sound basis of fact then we are part of the way towards our goal of protecting this immensely valuable wader area.

Just as there is pressure on the Wadden Sea, so in many parts of the British Isles the wader habitat is threatened. The review of British estuaries (Evans) highlights one of the areas of wader biology we know least about but which is of great importance: the movement during the winter of birds between estuaries. As research proceeds, it becomes more and more clear that our estuaries are not disparate areas which isolate groups of birds for long periods. They are a network of refuges some of which are important at one time of the year and others at another. The whole forms a complex system and the loss of any part of it could have large effects on others. The system extends far outside western Europe, to Morocco (Kersten & Smit) and Mauritania and the latter clearly is of more importance than initial studies indicated (Engelmoer, Piersma, Altenburg & Mes). We need to know urgently how and when waders move between the different areas of wader concentrations. We are just beginning to accumulate this information but it is critical that we find out to what extent birds settle for the winter in specific areas. It could well be that the small estuaries house only transitory populations in that there is sufficient food only on the larger areas to ensure a prolonged stay. In this context the marking of birds on small estuaries may well prove more fruitful than marking birds on larger estuaries as at least their emigration may be more easily recorded.

Whilst ensuring the protection of their habitat, it might be recognized that waders are highly adaptable birds and that during the course of their history, they have overcome drastic changes to their habitat, particularly during the past 20 000 years (Hale). The fact that a particular area has a specific role to play in the biology of wading birds at the present time does not mean that it will have this use indefinitely. Changes will occur and these must be monitored; waders can cope with the natural changes in the environment but may need help to survive the depredations of *Homo sapiens*. The population of *H. sapiens* in the UK alone is eight times that of the total waders using the so-called East Atlantic flyway; waders can hardly expect to be unaffected by the sheer size of the known human population but wader research can go a long way to reducing this effect. The more we know, the better able we will be to combat it.

Reference

Hale, W. G. (1980). *Waders*. Collins, London.

12

The Danish Wadden Sea*

K. LAURSEN & J. FRIKKE

Introduction

The Wadden Sea stretches from Den Helder, Netherlands, in the south, to Blåvand Huk, Denmark, in the north and the Danish Wadden Sea is situated between the west coast of Jutland and a chain of four islands running southwards from the peninsula of Skallingen. This basin between the mainland and the islands is strongly influenced by the tide which, on average, has an amplitude of about 2 m. The area of the Danish Wadden Sea is 850 km² and about 60% is formed of flats which are normally covered during high tide. Of the 500 km² of flats, mudflats cover 50 km², sand mixed with mudflats 100 km² and sandflats 350 km². Of the sandflats, some areas which are normally exposed during high tide, highsands, make up 10 km² (calculated from Smidt 1951).

Because of the northerly position of the Danish Wadden Sea it is the last major area in which many birds forage during spring migration. Correspondingly during autumn migration it is an important place at which to replenish fat reserves. The northerly position of the Danish Wadden Sea also means that the climate is somewhat harsher than over the rest of the Wadden Sea. The mean temperature in January is about 0°C and the mean date of the first frost in autumn is 23 October (Lysgård 1968), compared with the mean temperature in January of 2°C in the Netherlands (Mayer 1974).

Birds in the Danish Wadden Sea

Knowledge of the numbers of birds using the Danish Wadden Sea is based mainly on two investigations. In 1965–73 Joensen (1974) made a survey of the non-breeding populations of ducks, swans and Coot in Denmark and in 1978 Meltofte (1980) took a census of the

*Communication No. 187 from Vildtbiologisk Station, Kalø, 8410 Rønde, Denmark.

waders in the Danish Wadden Sea by aerial surveys supported by ground counts. From these two studies we have reliable information on the numbers and the seasons at which the different species within these groups are present. The Brent Goose (*Branta bernicla*) has also been studied in the Danish Wadden Sea (Fog 1972).

Besides these studies many reports from smaller parts of the Wadden Sea have been published – for example, the reports from Jordsand (Jepsen 1975) and Højer (Gram 1981).

In Table 12.1, the numbers of Brent Geese, dabbling ducks and waders present in each month of the year are summarized. For Brent Geese and the dabbling ducks the census was carried out for several years and for these species a monthly mean has been calculated. (Brent Geese were counted during 1966–71 and dabbling ducks during 1965–73.) The census of waders has been carried out for one year only.

It appears from the table that only small numbers spend the winter in the Danish Wadden Sea. This is especially pronounced for the waders: only 4 % of the maximum number are present in January and February. From Table 12.1 it can be seen that these are chiefly Oystercatchers (*Haematopus oestralegus*). For dabbling ducks the proportion is higher, about 30 % of the maximum number overwinter, chiefly Mallard (*Anas platyrhynchos*).

For those wildfowl species where the census was carried out over several years, Figs. 12.1 and 12.2 show the mean, the maximum and the minimum numbers recorded in each month. These refer to the most important dabbling duck species, the Shelduck (*Tadorna tadorna*), and the Brent Goose. The figures indicate the degree to which each species used the area in each month. The numbers of most species, including the Brent Goose, vary considerably from year to year during the period November to March. The Mallard and sometimes also the Shelduck may be present in large numbers in November and December. These pronounced annual fluctuations may be due to variation in the winter climate, a consequence of the northerly position of Danish Wadden Sea.

Diving ducks are also present offshore in the Danish Wadden Sea but the seasonal variation in numbers of these species is difficult to determine from the data available at present. The Common Scoter (*Melanitta nigra*) is present in numbers from a few thousand to 35 000 in November and from a few thousand to about 75 000 in January. Similarly the number of Eider (*Somateria mollissima*) is 15 000–30 000 in November, 35 000–65 000 in January and up to 50 000 in March. The Long-tailed Duck (*Clangula hyemalis*), Goldeneye (*Bucephala*

Table 12.1. Numbers of some bird species in the Danish Wadden Sea. For Brent Goose, dabbling ducks and Shelduck a mean number for several years has been calculated

	Jan.	Feb.	March	April	May	June	July	Aug.	Sept.	Oct.	Nov.	Dec.
Brent Goose (B. bernicla)	700	750	850	3000	1750	—	—	—	550	3700	3500	1350
Shelduck (T. tadorna)	7300	9700	10800	1500	1000	—	1000	500	16500	18700	6500	10000
Mallard (A. platyrhynchos)	15200	11000	13000	10000	250	—	300	500	8000	6700	22750	20300
Teal (A. crecca)			250	3000				1000		6300	10300	2300
Wigeon (A. penelope)	500	1300	4500	4500					14000	37000	17700	3700
Pintail (A. acuta)		300	250		—				1500	3300	2000	300
Dabbling ducks including Shelduck (total)	23000	22300	28800	19000	1250	—	1300	2000	40500	72300	59250	36600
Oystercatcher (H. ostralegus)	21900	14000	30100	25200	15600	4620	4970	28400	65000	51300	34500	44000
Lapwing (V. vanellus)			4610	658	724	846	805	9070	4730	6660	2450	27
Ringed Plover (C. hiaticula)			165	258	200	245	288	413	259	18	2	
Kentish Plover (C. alexandrinus)					3	17	36	16	16			
Golden Plover (P. apricaria)	450		833	2000	2790	15	21	7110	3060	2120	5280	125
Grey Plover (P. squatarola)	6	14	13	66	2950	474	28	2170	1590	1280	2320	15
Turnstone (A. interpres)			1		163	56	11	46	293	50	125	
Snipe (G. gallinago)	58		4	4	3	7		40	48	149	124	18
Curlew (N. arquata)	558	13	1100	1510	116	64	1330	3550	2060	3520	2130	4510
Whimbrel (N. phaeopus)				1	16	14	135	130				
(Numenius sp.)							100	100				
Black-tailed Godwit (L. limosa)				59	25	94	46	10	1		1	
Bar-tailed Godwit (L. lapponica)	4		2890	22500	42400	1100	6480	20300	13900	450	148	585
Common Sandpiper (T. hypoleucos)				1	49	1	36	105	15	1		
Green Sandpiper (T. ochropus)								4				
Wood Sandpiper (T. glareola)				2			1	2				
Redshank (T. totanus)	983	724	729	1640	2040	900	3740	10300	2880	2520	1470	1859
Spotted Sandpiper (T. erythropus)					24	6	5	72	33	55	2	
Greenshank (T. nebularia)				1	566	11	970	1750	570	88	5	
Knot (C. canutus)			5030	36	1	175	3	1300	1360	409	270	300
Little Stint (C. minuta)								18	411	6		
Temminck's Stint (C. temminckii)				1	1							
Dunlin (C. alpina)	4520	983	39200	196000	140000	705	5900	235000	358000	147000	63500	11100
Curlew Sandpiper (C. ferruginea)								56	20			
Sanderling (C. alba)			30	30	29	9	10	23	120	20	100	24
Ruff (P. pugnax)				51					8			
Avocet (R. avosetta)			104	797	635	654	2230	6390	3660	759		
Waders (rounded total)	28500	15700	84800	250800	208000	10000	27100	326400	458000	216400	112400	62600

'—' means no census taken.
After Fog (1972), Joensen (1974) and Meltofte (1980).

Fig. 12.1. Top: mean number of dabbling ducks including Shelduck in the Danish Wadden Sea 1965–75. Bottom: number of waders in 1978. '÷' indicates no data.

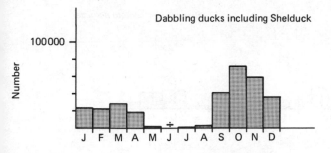

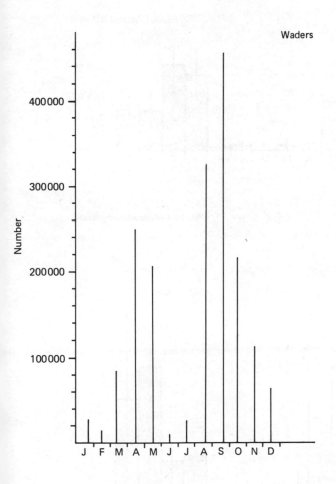

Fig. 12.2. (*a*) and (*b*). Mean number of Brent Goose, Shelduck, Mallard, Teal, Wigeon and Pintail together with the maximum and minimum in the different months.

(*a*)

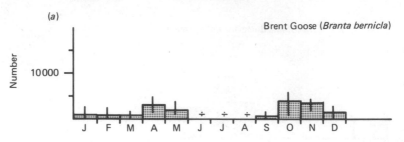

Brent Goose (*Branta bernicla*)

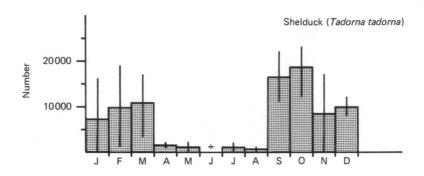

Shelduck (*Tadorna tadorna*)

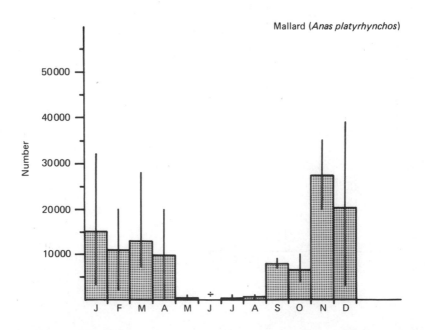

Mallard (*Anas platyrhynchos*)

(b)

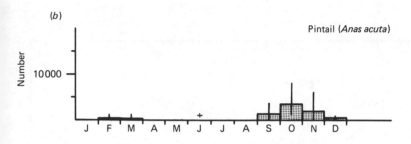

Pintail (*Anas acuta*)

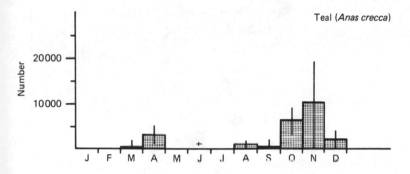

Teal (*Anas crecca*)

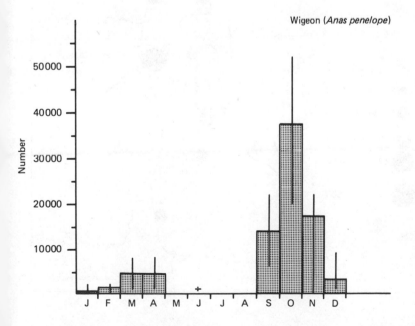

Wigeon (*Anas penelope*)

Fig. 12.3. (*a*) Mean number of waders on high-tide roosting places and (*b*) on the flats in August and September 1980.

(*a*)

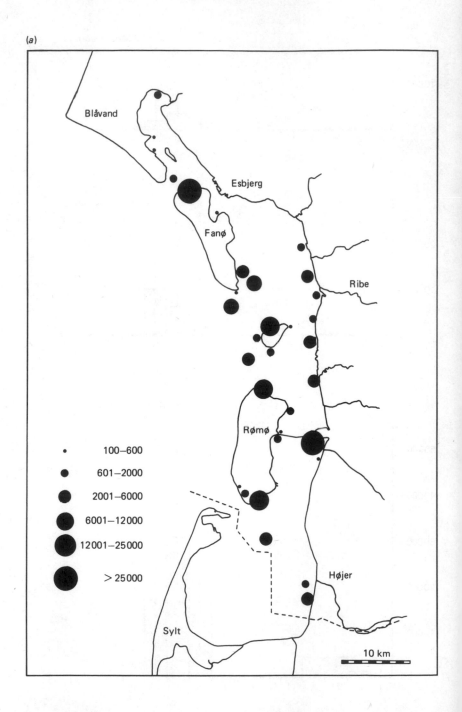

(*b*)

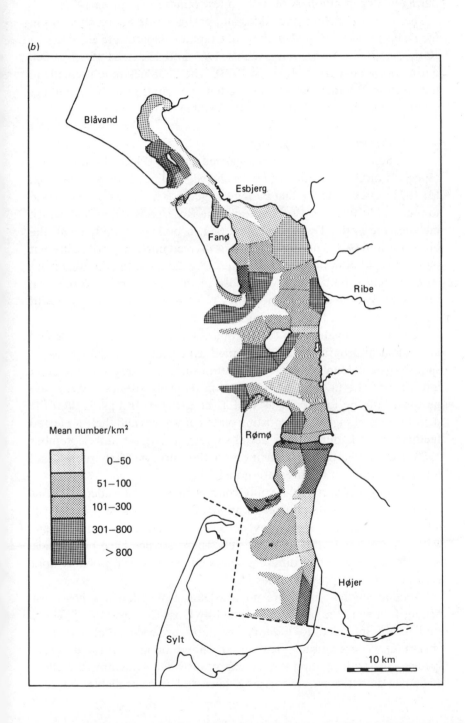

clangula) and Red-breasted Merganser (*Mergus serrator*) are present during winter in numbers of up to a few hundred (Joensen 1974).

The Common Scoter, the Eider and probably the Shelduck moult in the Danish Wadden Sea. Numbers of Common Scoter were estimated at 100 000–150 000 during July–August 1963 and Eider at about 30 000 during the same period (Joensen 1973). The Shelduck moults chiefly in the German Wadden Sea and only a few thousand, probably moulting, occur during summer in the Danish Wadden Sea.

Present and future projects

Some studies have been carried out in Danish Wadden Sea, especially on the migration of ducks and geese (Fog, J. 1968, 1974; Fog, M. 1972) and on the feeding biology of geese (Gram 1979). In relation to the establishment in 1979 of the 'Game Reserve Wadden Sea', which includes the entire Danish Wadden Sea, a biological study of all the birds was initiated. The game reserve is divided into zones with different types of regulation of human activities, e.g. zones with admittance but with hunting prohibited. The purpose of the new study is to describe how birds use the area and how they react to the different degrees of disturbances in the different zones.

The study is divided into two sections: a monitoring programme and an ecological programme. In the monitoring programme all species of shorebirds are counted by monthly aerial surveys. Because of the great influence of the tide on the distribution of many species, surveys are necessary during low water as well as high water. In Fig. 12.3*a,b* the distributions of high water roosting places of waders and their preferred feeding grounds during four aerial surveys in August and September 1980 are summarized. In addition to the surveys of shorebirds, all human activities in the area are noted.

The ecological part of the study concentrates on the waders. Special interest is attached to the distribution of the species on the flats in relation to sediment type, prey density, the density of the different wader species and other bird species. These studies have been made along transects covering different sediment types which probably hold different levels of prey densities.

Foraging studies of waders on the Danish Wadden Sea have just begun, following the methods outlined by Goss-Custard (1970). These studies are intended to identify the food preferences of the most important waders and to calculate the energy budgets of the different species. Such information is very important when evaluating the effect of human activities on the foraging times and areas available to birds.

From these studies we hope to be able to produce answers to some of the following questions. How many birds and of what species rest where and for how long in the Danish Wadden Sea? How many pass through the area? Where do they come from and where do they go to? Which factors regulate the numbers present: food densities, the density of the birds, the winter climate (already mentioned) or recreational human activities? These answers are very important for the future planning of the Danish Wadden Sea.

References

Fog, J. (1968). Dispersal of Teal (*Anas crecca*) on feeding flights from a Danish resting place – the Albue Bugten Refuge on the Island of Fanø. *Dansk Ornith. Foren. Tidsskr.* **62:** 32–6. (English summary.)

Fog, J. (1974). Gråand. In *Nyt dansk Jagtleksikon*, vol. 3, ed. A. Weitmeyer & F. Hansen, pp. 784–803. Brenner & Korch, København.

Fog, M. (1972). *Status for Knortegåsen (Branta bernicla)*. Vildtbiologisk Station, Kalø, 8410 Rønde, Denmark. 105 pp.

Goss-Custard, J. D. (1970). The responses of Redshank (*Tringa totanus* (L.)) to spatial variation in the density of their prey. *J. anim. Ecol.* **39:** 91–113.

Gram, J. (1979). *Forekomsten af gæs i Tøndermarsken*. Fredningsstyrelsen, Miljøministeriet, København. 75 pp.

Gram, J. (1981). *Ornithologisk undersøgelser i Tøndermarsken*. Fredningsstyrelsen, Miljøministeriet, København. 35 pp.

Jepsen, P. U. (1975). *Game Reserve 'Vadehavet' with the island of Jordsand*. *Danske Vildtundersøgelser* **24.** 80 pp. (English summary.)

Joensen, A. H. (1973). *Moult migration and wing-feather moult of seaducks in Denmark*. *Danish Rev. Game Biol.* **8,** No. 4. 42 pp.

Joensen, A. H. (1974). *Waterfowl populations in Denmark 1965–1973*. *Danish Rev. Game Biol.* **9,** No. 1. 206 pp.

Lysgård, L. (1968). Danmarks klima. In *Danmarks Natur*, vol. 2, ed. T. W. Böcher, A. Schou & H. Volsøe, pp. 112–28. Politikens Forlag, København.

Mayer, F. (1974). *Dierckes Welt Atlas*. Georg Westermann Verlag, . Braunschwieg.

Meltofte, H. (1980). *Wader counts in the Danish part of the Waddensea 1974–1978*. Fredningsstyrelsen, 50 pp. (English summary.)

Smidt E. L. B. (1951). *Animal production in the Danish Waddensea*. *Medd. Komm. Danmarks Fiskeri-og Havunders.*, Ser. Fiskeri, **11**(6).

13

The German Wadden Sea

P. PROKOSCH

Introduction

Within the total Wadden Sea area along the North Sea coast of the Netherlands, the Federal Republic of Germany and Denmark, the German section forms the largest part. Covering 4500 km^2, it represents 62% of this largest continuous tidal area on the flyways of Palaearctic waders. Two-thirds consists of eulittoral mud and sandy flats. The rest is composed of sub-littoral tide streams, rivers or creeks and (to a smaller extent) of supralittoral sands and saltings. Not included are the dune islands of Sylt, Amrum and the barrier chain of the seven East-Frisian islands between Borkum and Wangerooge as well as the embanked marsh islands of Nordstrand, Pellworm, Neuwerk and Föhr. Other small islands are the 10 so-called 'Halligen' (unembanked marshes with settlements only on tiny artificial hills) and seven flat vegetated sands which still belong to the upper supralittoral zone. They are regularly flooded by the sea in storm-tide situations with high water levels, 0.5 to 1.5 m above normal. Therefore, they form either isolated saltings (Halligen) or dune vegetation zones (on the sandy substrates).

Two parts of the German Wadden Sea can be distinguished if we take the political boundary – the River Elbe – of the two federal states Schleswig–Holstein and Niedersachsen (Lower Saxony) as a criterion. For some later aspects it seems sensible to follow this division as former descriptions of Wadden Sea shorebirds (Smit & Wolff 1980) have kept to this or have dealt with the Schleswig–Holstein Wadden Sea only (Drenckhahn, Heldt & Heldt 1971; Busche 1980). The Niedersachsen Wadden Sea is the subject of a work in progress (Goethe, Heckenroth & Schumann 1978) on the birds of the whole of Lower Saxony.

From the ecological/geomorphological point of view other distinctions are more evident. There is, for example, the deep bight of the

Dollart (mouth of the River Ems) on the Dutch/German border with very much its own characteristics (Dahl & Heckenroth 1978*a*) and Jade Bay which is a relic of the last great advances of the sea (storm tides during the last centuries) into the already reclaimed coastal marshland (Reineck 1970). Apart from these bights the German Wadden Sea can be divided into three main sections.

1. The North-Frisian Wadden Sea. Situated between the Eiderstedt peninsula and the Danish/German border this area shows the greatest variety of different habitats. Recently consideration was given to its becoming a national park (Erz 1972). The variability of the many islands (two dune islands, two marsh islands, one mixed Pleistocene/marsh island, ten Halligen and three huge unvegetated high sands) is one important aspect of the area; the other (more important for the waders) is the enormous extent of the tidal flats. Over a distance of up to 40 km between the outer sand barriers and the mainland the sea bottom is visible (and for the waders reachable) during the two low-tide phases each day. The irregular coastline with small bights and bars to some of the islands leads to many sheltered areas where sedimentation can take place easily and highly productive mud- or mixed (mud/sand) flats are formed (Table 13.1). As the end point of such sedimentation processes saltmarshes arise on a coastline above the normal high water level. Today the growth of most of the North-Frisian saltings has been encouraged through artificial reclamation work. Including the Halligen the saltmarsh habitat consists of 6200 ha (1981), 3.5% of the total North-Frisian Wadden Sea area (Fig. 13.1).

2. The Inner German Bight (Wadden Sea between Eider and Jade). This central part of the German Wadden Sea is very much affected by the two estuaries of the Rivers Elbe and Weser and – to a lesser extent – of the Rivers Eider and Jade. Their freshwater input lowers the salinity of the North Sea water from 29‰ in the outer Wadden Sea zone to about 10‰ in the mouths of the Rivers Elbe and Weser (compare 30–31‰ in the North-Frisian and East-Frisian Wadden Sea with 34–35‰ in the central North Sea). These rivers also play a role in the extremely high dynamic geomorphological processes of this region. A sand bar system with higher sand and dune islands (Mellum, Knecht-sand, Scharhörn, Trischen, Blauort and Tertius) is changing continuously (Dijkema, Reineck & Wolff 1980). Only the one marsh island (Neuwerk) keeps its position fairly stable. The relative amount of coarse-grained tidal surface sediment (sandy flats) compared with fine sediments (mixed flats and mudflats) is considerably higher than, say, in

Fig. 13.1. The Schleswig–Holstein Wadden Sea.

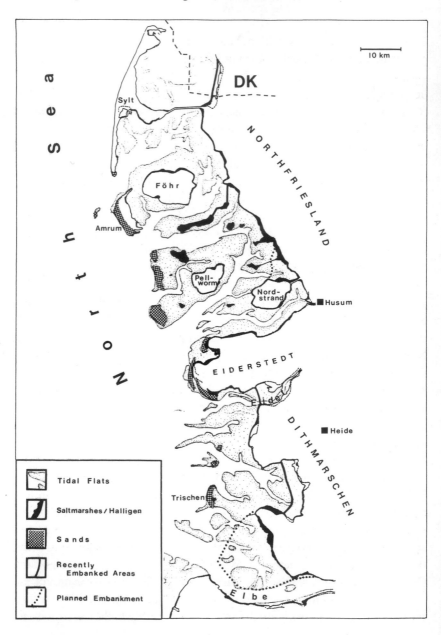

Table 13.1. *The tidal flat sizes of the German Wadden Sea sections, differentiated into three general surface sediment types, given in km², and (in parenthesis) the percentage of each sediment.*

Sediment of flats	North Friesland	Inner German Bay	East Friesland	Jade Bay	Dollart	Schleswig–Holstein	Nieder-sachsen	Total German Wadden Sea
Sand	650 (64)	1060 (79)	405 (65)	35 (29)	12 (13)	1160 (73)	1002 (63)	2162 (68)
Mixed sand/mud	310 (31)	220 (16)	140 (22)	25 (21)	20 (22)	375 (23)	340 (21)	715 (22)
Mud	50 (5)	65 (5)	80 (13)	60 (50)	58 (65)	65 (4)	248 (16)	313 (10)
Total	1010	1345	625	120	90	1600	1590	3190

After MELF 1981, P. Wieland, personal communication, Ragutzki 1981 and own planimetric records.

the North-Frisian Wadden Sea section (Table 13.1; Figs. 13.1, 13.2). 3. The East-Frisian Wadden Sea. The morphology of this western section of the German Wadden Sea (between the Dutch border and the River Jade) is comparable with the Dutch Wadden Sea. With its typical chain of large barrier islands, its tidal area as a long strip between them and the more or less straight coastal sea wall, it represents a continuation of the Wadden Sea along the West-Frisian islands in the Netherlands (Dijkema *et al.* 1980). As the only major bight on the mainland coast the Ley Bay has remained and forms with its soft mudflats and expanded saltings an important wader feeding and roosting area (Dahl & Heckenroth 1978*b*). Similar to the North-Frisian Wadden Sea, the East-Frisian tidal flats consist of about 35% of fine sediments (mixed flats and mudflats; Table 13.1). Suitable high water roosting places for waders can be found on the mainland strip of saltmarshes, three small island (Memmert, Lütje Hörn and Minsener Oldeoog) and in particular on the sandy hooks and saltings on the southeast ends of the seven big islands (Fig. 13.2).

The birds

Relatively little is known about the waders of the German Wadden Sea (ct. research work in the Netherlands and Britain). Until recently in Lower Saxony we had not even succeeded in getting complete simultaneous surveys of the whole area. For this reason it is likely that former publications on waders of the German Wadden Sea (e.g. Prater 1976; Smit & Wolff 1981) underestimated the total numbers.

In 1980 a new attempt was made to cover the shorebirds of the whole Wadden Sea. Together with Netherlands and Denmark, six synchronized counts on selected dates throughout the year were carried out between April 1980 and September 1981. In Germany we had the opportunity to combine a ground and aerial survey in Schleswig–Holstein and Niedersachsen (Knief 1983). On three of these counts nearly full coverage was obtained and the numbers of waders counted are given in Table 13.2.

Taking into account the general problems of counting such large numbers of waders (Kersten, Rappoldt & Smit 1981), these figures present minimum numbers by any standards. Adding the highest counts of each (Table 13.3), at least two million wading birds use the coasts of Schleswig–Holstein (1.1 million) and Niedersachsen (0.9 million) every year. This does not take into account the turnover of different individual birds during the migration seasons. Because there is a continuous flow

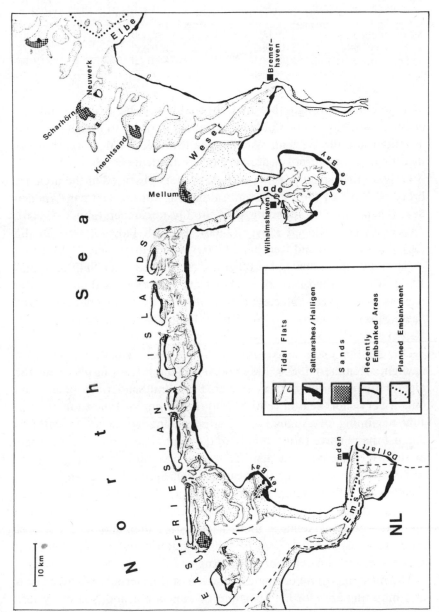

Fig. 13.2. The Niedersachsen Wadden Sea.

Table 13.2. *Total numbers of waders in the German Wadden Sea counted in 1980–81 during three surveys with maximum coverage*

	Schleswig–Holstein	Niedersachen	Total
8 November 1980	470 000	730 000	1 200 000
17 January 1981	110 000	270 000	380 000
26 September 1981	810 000	600 000	1 410 000

of individuals through the area, it could well be that the autumn numbers will have to be doubled or trebled in order to obtain the total actually using the German Wadden Sea (Smit & Wolff, 1981). Further investigations on turnover are very much needed.

Only a fairly small portion of these waders observed in the area are breeding birds of the islands and saltmarshes of the German Wadden Sea. There are the breeding pairs of Oystercatcher (about 11 000), Avocet (3000), Ringed Plover (1400), Kentish Plover (900), Dunlin (30), Curlew (120) and Redshank (7000), which total some 24 000 pairs, 48 000 adult individuals plus offspring (simplified after Smit & Wolff, 1981). Most of the birds are visitors from the boreal and arctic tundra regions in the west Palaearctic and (in case of Turnstone, Knot and Sanderling) the northeast Nearctic (Glutz von Blotzheim, Bauer & Bezzel 1975/77).

Due to the extreme conditions in the north (food, temperature, daylight; Remmert 1980), the majority of the birds – in particular the high-arctic ones such as Grey Plover, Knot and Sanderling – can spend only a very short season (end of May/beginning of June to the end of July/beginning of August) on their breeding grounds. As early as August the massive return builds up the annual peak of wader numbers in the German Wadden Sea. The very first birds returning are the 'emancipated' females of the Scandinavian Spotted Redshanks. After they have left their clutches to their males (Hilden 1961), they arrive in the second half of June (Drenckhahn *et al.* 1971; Schultz, 1980). The Spotted Redshank females immediately start their moult as do most other waders on arrival. They have light winter plumage when the black males and dark juveniles arrive in August.

The most important ecological function of the German Wadden Sea is as a moulting area for high portions of Palaearctic and Nearctic wader populations (Drenckhahn *et al.* 1971). Only a few birds stop for short periods without moulting, i.e. Whimbrels and some Bar-tailed Godwits, Greenshanks, Siberian Knots and Curlew Sandpipers. Most of the others listed in Table 13.3 do stay for longer (until October) to complete

Table 13.3. *Numbers of waders in the two sections of the German Wadden Sea*

Species	Autumn maximum numbers			Winter numbers (January 1981)			Spring maximum numbers		
	S-H	NS	Total	S-H	NS	Total	S-H	NS	Total
Oystercatcher	150 000 (10)	310 000 (10)	460 000 (10)	65 000	170 000	235 000	40 000 (3/4)	175 000 (3)	215 000 (3)
Avocet	5 000 (7)	22 000 (9)	24 000 (9)[a]	0	200	200	2 000 (4/5)	2 000 (3/4)	4 000 (4)
Ringed Plover	10 000 (8)	5 000 (9)	15 000 (9)	0	200	200	4 000 (5)	4 000 (5)	8 000 (5)
Kentish Plover	2 000 (8)	1 000 (8)	3 000 (8)	0	0	0	500 (5)	500 (5)	1 000 (5)
Grey Plover	15 000 (8)	20 000 (8)	35 000 (8)	300	1 000	1 300	25 000 (5)	25 000 (5)	50 000 (5)
Turnstone	2 000(8)	1 000 (8)	3 000 (8)	300	800	1 100	1 000 (5)	1 000 (5)	2 000 (5)
Curlew	40 000 (9)	45 000 (10)	85 000 (10)	15 000	35 000	50 000	20 000 (3)	40 000 (3)	60 000 (3)
Whimbrel	2 000 (8)	2 000 (8)	4 000 (8)	0	0	0	500 (5)	500 (5)	1 000 (5)
Bar-tailed Godwit	50 000 (8)	50 000 (8)	100 000 (8)	3 000	3 000	6 000	90 000 (5)	100 000 (5)	190 000 (5)
Spotted Redshank	4 000 (8)	2 000 (8)	6 000 (8)	0	0	0	300 (5)	2 500 (5)	2 800 (5)
Redshank	16 000 (7)	15 000 (7)	31 000 (7)	1 500	4 000	5 500	10 000 (4/5)	10 000 (4/5)	20 000 (4/5)
Greenshank	3 000 (8)	2 000 (8)	5 000 (8)	0	0	0	300 (5)	200 (5)	500 (5)
Knot	400 000 (8/10)	200 000 (9)	500 000 (9)[a]	8 000	8 000	16 000	300 000 (5)	100 000 (4)	350 000 (5)[a]
Sanderling	4 000 (10)	15 000 (9)	19 000 (9)	100	500	600	17 000 (5)	11 000 (5)	28 000 (5)
Curlew Sandpiper	1 000 (8)	1 000 (8)	2 000 (8)	0	0	0	100 (5)	100 (5)	200 (5)
Dunlin	400 000 (8/10)	200 000 (9)	600 000 (9)	10 000	50 000	60 000	300 000 (5)	200 000 (5)	500 000 (5)

Schleswig–Holstein (S–H) and Niedersachsen (NS). For autumn and spring migration maximum figures are given and (in parentheses) the month when they occur. Winter numbers should represent an average January situation and are based mainly on a very complete count in January 1981 (moderate temperature). Other figures are combined from recent counts (1979–83), Busche 1980 and Smit & Wolff 1980. The early autumn (July/August) and the late spring (May) numbers of Niedersachsen are interpolations and estimates.

[a] Total is not the sum of S–H and NS as maxima were reached at different times of year, and partly the same birds were counted in both areas.

their moult (Schultz 1980 and unpublished data; Smit & Wolff 1981). The extra energy requirement of the moulting process coincides with the maximum standing crop of the invertebrate food resources (Drenck-hahn 1980).

Especially during October/November many waders suddenly leave the German Wadden Sea to continue their migration to the wintering grounds, which lie along the Atlantic coast from the British Isles in the north (Dunlin) to the south of Africa (Siberian Knot). Up to 400 000 waders remain in the area during mild winters (Table 13.3), declining to about 150 000 in extremely cold seasons (e.g. in January 1982, when 90 % of the tidal flats in Schleswig–Holstein and 80 % in Niedersachsen were covered with ice). In accordance with the temperature, 70–95 % of the wintering waders belong to the bigger species of waders, Oyster-catcher and Curlew (compare with 25–30 % in autumn, Table 13.3). It seems likely that the bigger birds can survive better than smaller ones because they lose relatively less body heat, have longer bills to reach deeper prey and have better adapted feeding strategies for coping with coldness (Pienkowski 1978/79; Drenckhahn 1980). Furthermore, long storm periods with high water levels even during low tides reduce the feeding time and space. This, in addition to ice covering the higher flats, affects the smaller waders more.

Significantly, more waders do stay through the winter in Niedersach-sen than in Schleswig–Holstein (Table 13.3). This trend of increasing winter numbers the further one goes southwest in the Wadden Sea can be followed as far as the Netherlands, where the mean water tempera-ture in the coldest time of year is 2–3 °C higher than in Schleswig–Holstein (Drenckhahn 1980).

The other main function of the German Wadden Sea seems to be its importance as a spring staging post for arctic waders. On their way north they stop here (mostly for a few weeks), to build up fat reserves. In the case of Dunlin the birds nearly double their lean weight. In mid-March, after having used up the winter reserves, Dunlins in the Wash in southeast England show minimum weights of only 45–50 g (Branson 1979), whereas in the last days of May the same birds in Schleswig–Holstein have a mean weight of 75–80 g and some individuals may reach 92 g (Schultz 1980).

Food
In contrast with the late summer situation, when maximum numbers of waders coincide with optimal food supply, the feeding resources reach their annual minimum in March and reach the mean annual biomass by May (Beukema 1974; Reise 1979).

About 70 % of the waders observed in autumn have to use the area again on their way back to the breeding grounds, the others having died, stayed as non-breeders further south, or used a different migration route. These birds are forced to obtain extra energy reserves to cope with long flight distances and the prospect of little food in the arctic at the time of their arrival. Having this in mind, the observed differences in passage times in Schleswig–Holstein (Busche 1980; Drenckhahn 1980) seem to be quite sensible (less interspecific competition): the early breeders of the boreal zone (Curlew, Oystercatcher, Ringed Plovers of the nominate form) peak in March/April and have mostly left by the end of April, whereas most other waders (in particular the high-arctic species) concentrate in May. The Knot demonstrates that even different populations of the same species occur at different times: when the Greenland/Canadian Knots are leaving the German Wadden Sea in mid-May, the Siberian ones (coming from Africa) are only just arriving. The latter have used their 80-g fat reserves they put on in Africa. During their stay of 2–3 weeks in the North-Frisian Wadden Sea and the Inner German Bay (it appears that nearly the whole Siberian population stays exclusively in this area during the second half of May/first week of June) they reach their maximum weight again. These reserves enable them to make the last part (about 4000 km) of their way to northern Siberia in an almost non-stop flight (Dick 1979; Dick, Fournier & Prokosch 1980; Schultz 1980 and unpublished data; Boere & Smit 1981).

Only a few local investigations exist (Ehlert 1964; Höfmann & Hoerschelmann 1969; Schultz 1980) concerning the food waders take in the German Wadden Sea. It may not be justifiable to generalize these results for the whole area and for all seasons of the year. Nevertheless it seems obvious that the following macrobenthic animals of the tidal flats do play important roles: Mollusca: *Cerastoderma* (Oystercatcher, Curlew), *Mytilus* (Oystercatcher, Curlew), *Mya* (Oystercatcher, Curlew), *Macoma* (Oystercatcher, Knot), *Littorina* (Oystercatcher, Curlew, Bartailed Godwit, Knot), *Hydrobia* (Dunlin, Knot, Avocet and to a lesser extent most other species); Annelida: *Nereis* (main prey of many waders, in particular Grey Plover, Avocet, Dunlin, *Tringa* spp.), *Arenicola* (Oystercatcher, Curlew, Bar-tailed Godwit); Crustacea: *Carcinus* (Oystercatcher, Curlew, *Tringa* spp.), *Crangon* (*Tringa* spp.), *Corophium* (Knot, Redshank, Dunlin). Small fish are regularly eaten by Greenshank and Spotted Redshank. Insects, mainly found on the supralittoral saltings and sands or on the tide edge, are food of the Sanderling, the small plovers (*Charadrius*) and to a small extent other waders, which feed during high water on their roosts.

All the mentioned macrobenthic food species belong to the 15 most frequent bigger invertebrates of the tidal flats (Michaelis 1969, 1976; Reise 1979; Meyer & Michaelis 1980; Drenckhahn 1980). The level of the flats (= time of water covering) probably affects the distribution of the macrobenthic fauna more than the granular size of the sediment (Reise 1979). Reise found in Nordstrand Bay in the summer of 1979 (after a severe winter) a mean biomass of 67.6 (g dry weight/m^2). The bigger animals (>0.2 cm^3 body volume) had their maximum (38.3 g/m^2) in the upper third of the tide range, the small individuals (<0.2 cm^3) in the lower third (54.5 g/m^2). Of the three different sediment types (Table 13.1) he recorded the highest biomass of both size classes in the mud (84.6 g/m^2), followed by the mixed flats (70.8 g/m^2), and the lowest in the sandy substrate (47 g/m^2).

Threats to the environment

This adds to the list of factors making the high flats of fine sediments close to the coast or islands exceptionally valuable for waders (see e.g. Evans & Dugan, Chapter 1 of this volume, Swennen, Chapter 10 of this volume). Therefore, it is of particular concern that these areas are the most threatened. In all parts of the German Wadden Sea there are reclamation plans for the high mudflat and saltmarsh habitat: in North Friesland the Rodenas-Vorland (550 ha of mostly saltmarsh) on the Danish border has been embanked in 1981. Owing to future saltmarsh reclamation in front of the new sea wall further losses of high flats can be expected. After extensive lobbying by conservationists the reclamation plan for Nordstrand Bay has been cut down from 5600 ha to 3300 ha (760 ha saltmarsh, 2400 ha tidal flats of mostly mixed or muddy sediments and 140 ha sublittoral streams). It seems to be likely that already in 1982 the roosting and feeding area of at least 80 000 coastal waders will be disappearing behind a new sea wall (Prokosch 1979; Schultz 1980; MELF 1981).

In the centre of the Inner German Bight there is still the 20 000 ha project of 'Medemland' in the mouth of the River Elbe under discussion (Heydemann 1981*a*). This is a combined industrial/harbour/recreation project which will result in enormous tidal habitat losses (mostly sandy flats and saltings). In East Friesland the former plans to embank the whole of Ley Bay (Dahl & Heckenroth 1978*b*) have been dropped, but smaller projects still threaten this important Avocet and Spotted Redshank site. Inside the muddy Dollart the planned enlargement of the Ems harbour of Emden will probably reduce the area (again especially important for moulting Avocets and Spotted Redshanks) by

1150 ha and the salinity of the already brackish water (Dahl & Heckenroth, 1978*a*).

Another environmental problem in the German Wadden Sea is the pollution – especially of the Inner German Bight – due to increasing industrialization along the Rivers Elbe, Weser, Ems and Jade (Podloucky 1980; Rat der Sachverständigen für Umweltfragen 1980). The oil harbour of Wilhelmshaven is a permanent potential source of oil contaminations. Heavy shipping traffic to Hamburg, Bremerhaven and the Kiel canal constitute a further danger and recently discovered oil resources under the sea near Trischen will probably lead to exploitation.

A third threat arises from increasing recreational activities (Heydemann 1981*b*). Even inside the nature reserve of the North-Frisian Wadden Sea a new harbour for leisure craft was set up two years ago just opposite the most important Knot roosting site at Westerhever (up to 200 000 Knots) on the west end of the Eiderstedt peninsula. Heavy tourism takes place in particular on Sylt, Amrum Föhr and the East-Friesian islands, with airfields on nearly all of them.

To keep the position of the German Wadden Sea as one of the most important (at certain times of the year it may be the most important) wader areas in the western Palaearctic and Africa, very strong protection measures must be taken. Probably the status of a national park for the whole area would help. Present talks with neighbouring Denmark and the Netherlands could even lead to international steps for better protection of the total Wadden Sea.

References

Beukema, J. J. (1974). Seasonal changes in the biomass of the macrobenthos of a tidal flat area in the Dutch Wadden Sea. *Neth. J. Sea Res.* **10**: 236–61.

Boere, G. C. & Smit, C. J. (1981). Knot, *Calidris canutus*. In *Birds of the Wadden Sea*, ed. C. Smit & W. Wolff, pp. 136–45. Balkema, Rotterdam.

Branson, N. J. B. A. (ed.) (1979). Wash Wader Ringing Group, Report 1977–78. Trinity College, Cambridge, England.

Busche, G. (1980). *Vogelbestände des Wattenmeeres von Schleswig–Holstein.* Kilda, Greven.

Dahl, H.-J. & Heckenroth, H. (1978*a*). *Landschaftspflegerisches Gutachten zur Emsumleitung durch den Dollart. Natursch. u. Landschaftspfl. in Niedersachsen* **6**.

Dahl, H.-J. & Heckenroth, H. (1978*b*). *Landschaftspflegerisches Gutachten zu geplanten Deichbaumßnahmen in der Leybucht. Natursch. u. Landschaftspfl. in Niedersachsen* **7**.

Dick, W. J. A. (1979). Results of the WSG project on the spring migration of Siberian Knot *Calidris canutus* 1979. *Waders Study Group Bull.* **27**: 8–13.

Dick, W. J. A., Fournier, O. & Prokosch, P. (1980). WSG project on the spring passage of Siberian Knot. *Wader Study Group Bull.* **28**: 15.

Dijkema, K. S., Reineck, H. E. & Wolff, W. (eds) (1980). *Geomorphology of the Wadden Sea area.* Report 1 of the Wadden Sea Working Group.

Drenckhahn, D. (1980). Nahrungsökologische Aspekte zum Vorkommen der Wat- und Wasservögel im schleswig–holsteinischen Wattenmeer. In

Vogelbestände des Wattenmeeres von Schleswig–Holstein, ed. G. Busche, pp. 119–30. Kilda, Greven.

Drenckhahn, D., Heldt, R., Jr. & Heldt, R., Sr (1971). Die Bedeutung der Nordseeküste Schleswig–Holsteins für einige eurasische Wat- und Wasservögel mit besonderer Berücksichtigung des Nordfriesischen Wattenmeeres. Natur. u. Landsch. 46: 338–46.

Ehlert, W. (1964). Zur Ökologie und Biologie der Ernährung einiger Limikolen-Arten. J. Orn. 105: 1–53.

Erz, W. (1972). Nationalpark Wattenmeer. Hamburg u. Berlin.

Glutz Von Blotzheim, U. N., Bauer, K. M. & Bezzel, E. (1975/1977). Handbuch der Vögel Mitteleuropas, vol. 6/7. Wiesbaden.

Goethe, F., Heckenroth, H. & Schumann, H. (1978). Die Vögel Niedersachsens. Naturschutz u. Landschaftspfl. in Niedersachsen, Sonderreihe B, 2(1).

Heydemann, B. (1981a). Biological consequences of diking on saltmarshes and mud or sandy flats. In Environmental problems of the Waddensea-Region, ed. S. Tougaard & C. Helweg Ovesen, pp. 31–71. Fredningsstyrelsen/Fiskeri-og Søfartsmuseet Saltvandsakvariet, Esbjerg.

Heydemann, B. (1981b). Ökologie und Schutz des Wattenmeeres. Schriftenr. Bundesmin. f. Ern., Landw. u. Forsten, Angewandte Wissenschaft 255.

Hilden, O. (1961). Über den Beginn des Wegzuges bei den Limikolen in Finnland. Ornis Fenn. 38: 2–31.

Höfmann, H. & Hoerschelmann, H. (1969). Nahrungsuntersuchungen bei Limikolen durch Mageninhaltsanalysen. Corax 3: 7–22.

Kersten, M., Rappoldt, K. & Smit, C. (1981). Over de nauwkeurigheid van wadvogeltellingen. Limosa 54: 37–46.

Knief, W. (1983). Wat- und Wasservogelzählungen in Niedersachsen und an der schleswig–holsteinischen Westküste. Beiheft Natursch. u. Landschaftspfl. in Niedersachsen.

MELF (Minister für Ernährung, Landwirtschaft und Forsten Schleswig–Holstein) (1981). Gutachten zur geplanten Vordeichung der Nordstrander Bucht. Schriftenr. d. Landesreg. Schleswig–Holstein 12.

Meyer, M. & Michaelis, H. (1980). Makrobenthos des 'Hohen Weges' Jber. Forsch.-Stelle f. Insel- u. Küstenschutz Norderney 1979: 31.

Michaelis, H. (1969). Makrofauna und Vegetation der Knechtsandwatten. Jber. Forsch.-Stelle f. Insel- u. Küstenschutz Norderney 1967: 19.

Michaelis, H. (1976). Die Makrofauna des nördlichen Eversandes (Wesermündung). Jber. Forsch.-Stelle f. Insel- u. Küstenschutz Norderney 1975: 27.

Pienkowski, M. W. (1978/79). Differences in habitat requirements and distribution patterns of plovers and sandpipers as investigated by studies of feeding behaviour. Ver. orn. Ges. Bayern 23: 105–24.

Podloucky, R. (1980). Der Niederelberaum – Industrie kontra Natur. Elmshorn, Arbeitsgem. Umweltplanung Niederelbe (AUN).

Prater, A. J. (1976). The distribution of coastal waders in Europe and Africa. In Proceedings of the international conference on the conservation of wetlands and waterfowl, ed. M. Smart, pp. 255–71. Int. Waterfowl Res. Bureau, Slimbridge.

Prokosch, P. (1979). Waders in the Nordstrand Bay (Schleswig–Holstein) 1979. Wader Study Group Bull. 27: 4.

Ragutzki, G. (1981). Verteilung der Oberflächensedimente auf den niedersächsischen Watten. Report with maps of Forschungs stelle für Insel- und Küstenschutz, D-2982 Norderney.

Rat der Sachverständigen für Umweltfragen (1980): Umweltprobleme der Nordsee. Sondergutachten Juni 1980. Kohlhammer, Stuttgart u. Mainz.

Reineck, H.-E. (ed.) (1970). Dast Watt- Ablagerungs- und Lebensraum. Kramer, Frankfurt am Main.

Reise, K. (1979). Forschungsvorhaben zur Bodenfauna im Gebiet der

Nordstrander Bucht. Gutachten f. Landesreg. Schleswig–Holstein. (Short form published in MELF 1981.)

Remmert, H. (1980). *Arctic animal ecology.* Springer, Berlin, Heidelberg, New York.

Schultz, W. (1980). *Vogelkudliche Bedeutung der Nordstrander Bucht. Gutachten f. Landesreg. Schleswig–Holstein.* (Short form published in MELF 1981.)

Smit, C. & Wolff, W. (eds) (1981): *Birds of the Wadden Sea.* Balkema, Rotterdam. 308 pp.

14

The Dutch Wadden Sea

W. J. WOLFF & C. J. SMIT

Introduction

This chapter rests heavily on the information presented by several authors in the book *Birds of the Wadden Sea* (Smit & Wolff 1981). That book describes breeding areas, migration routes, wintering areas, moulting areas, annual cycles, population sizes, numbers per sub-area of the Wadden Sea, food composition, feeding activities and total food consumption for 32 bird species occurring abundantly in the Wadden Sea. The species concerned are Spoonbill, 2 species of geese, 8 species of ducks, 15 species of waders, 3 species of gulls and 3 species of terns. The book also contains chapters on habitat selection and competition in estuarine birds, the importance of the Wadden Sea for birds, production and consumption of bird food, and threats to the birds of the Wadden Sea. For details of the data presented in the following paragraphs the reader is referred to that volume.

The Wadden Sea is a shallow coastal sea of about 8000 km², which is situated behind a long chain of barrier islands along the coasts of the Netherlands, the Federal Republic of Germany and Denmark (Fig. 14.1). The Dutch part is called Waddenzee, the German part Wattenmeer and the Danish area Vadehavet. This is, in fact, the same name but in three different languages. In the international scientific literature the name Wadden Sea is used most often for the total area as well as for any of its parts.

The Dutch Wadden Sea measures about 2600 km², the exact area depending on the position where the boundaries are drawn. It comprises about 1300 km² of tidal flats, called 'wadden' (plural; singular, 'wad') and an equally large area of tidal channels and shallow water never becoming dry at low tide. The tidal range varies between 136 cm at Den Helder and 280 cm at Delfzijl. A series of barrier islands separates the

Dutch Wadden Sea from the North Sea. The five largest islands are inhabited permanently by man. At present there are five uninhabited barrier islands, two of them with a core of larger dunes and thus staying dry at all tides (Rottumerplaat, Rottumeroog). The other three uninhabited barrier islands, as well as some temporary islets in the outer deltas of the tidal inlets, are hardly more than high sandflats. At most they show low primary dunes which are flooded rather frequently. All islands are important for birds either as breeding areas or as high-tide roosts or both.

Fig. 14.1. The Wadden Sea, a shallow coastal sea along the coasts of the Netherlands, the Federal Republic of Germany and Denmark.

The tidal flats of the Dutch Wadden Sea are rather sandy (Fig. 14.2); muddy areas occur only in sheltered places. Also the occurrence of saltmarshes is dependent on shelter from wave exposure; altogether about 87 km² of marshes are left after centuries of reclamation.

Breeding birds

Table 14.1 summarizes the data on waders and waterfowl breeding in the Dutch Wadden Sea area. The numbers of Spoonbills, Avocets and Sandwich Terns are internationally important.

The breeding population of the Dutch Wadden Sea area does not have a constant size. Swennen (1982) shows how several species had very different population numbers in the past 80 years. Gulls and Eider Ducks showed a spectacular increase during this period (Fig. 14.3). This is at least partly due to better protection at the breeding places. Other species, such as terns and plovers, showed a decrease of their breeding populations.

Around 1965 a strong decline of many breeding species could be traced back to the effects of discharge of polluted waste water of a pesticide factory on the Rhine estuary near Rotterdam (Koeman 1971). The decrease of Sandwich Tern was especially dramatic (Fig. 14.4). After the factory was closed most of the bird species concerned have been increasing again year after year (Fig. 14.4).

Small populations of species not breeding in the Dutch Wadden Sea spend the summer in the area, e.g. Bar-tailed Godwit and Knot.

Fig. 14.2. Generalized map of the sediment composition in the Dutch Wadden Sea.

LEGEND

SAND (< 5% clay)

MUDDY SAND (5-12% clay)

MUD (>12% clay)

Autumn migration

Several of the species breeding in the Dutch Wadden Sea area leave in the summer and/or autumn. For example, Shelducks leave the breeding area to moult in the German Wadden Sea in July–August. Avocets and Kentish Plovers are hardly seen after November and September, respectively. Terns mostly disappear in August–September. In other species, such as the Ringed Plover, it is hard to distinguish between the breeding population of the Dutch Wadden Sea and migrants from more northerly regions, so it becomes difficult to ascertain when the breeding population has left.

A similar problem occurs with the species which do not breed in the Wadden Sea. In many areas we can distinguish between different groups of birds which constitute summering, migrating – sometimes from different origins – and wintering populations but the positioning of the boundaries between the Wadden Sea populations is very difficult. When a strong increase in the numbers of birds present is considered as an

Fig. 14.3. Population development of Eider, Lesser Black-backed Gull, Herring Gull and Common Gull in the Netherlands in the twentieth century. After Swennen (1982).

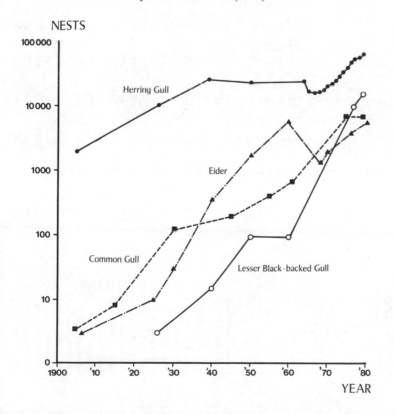

Table 14.1. Approximate numbers of breeding pairs of some bird species in the Dutch Wadden Sea area

	Texel (1974 with some data from later years)	Vlieland (1966 and later years)	Richel (1979)	Griend (maximal numbers 1979–81)	Terschelling (1977)	Ameland (1972)	Engelsmanplaat (max. 1977–79)	Schiermonnikoog (1973)	Rottumerplaat (max. 1977–79)	Rottumeroog (max. 1977–79)	Mainland Noord-Holland (max. 1977–79)	Mainland Friesland (partly 1972–74, partly 1979–81)	Mainland Groningen, excl. Dollard (1976)	Dollard (max. 1969–72)	Estimated size of total population
Platalea leucorodia	47	—	—	—	15	—	—	—	—	—	—	—	—	—	62
Anas platyrhynchos	1350	175	—	10	370	120	—	275	3	3	1	60	—	150	2517
Anas crecca	30	40	—	—	25	20	—	32	—	—	—	—	—	5	152
Anas querquedula	7	10	—	—	4	20	—	7	—	—	—	3	—	3	54
Anas strepera	2	4	—	—	7	1	—	5	—	—	—	—	—	2	21
Anas acuta	2	2	—	—	—	2	—	1	1	1	—	1	—	—	10
Anas clypeata	86	45	—	—	45	35	—	53	—	—	—	20	—	70	354
Aythya fuligula	160	4	—	—	16	35	—	25	—	—	—	7	—	—	247
Aythya ferina	2	—	—	—	1	5	—	2	—	—	—	—	—	—	10
Somateria mollissima	92	1800	—	1	900	90	—	700	20	41	—	—	—	—	3644
Tadorna tadorna	200+	100	—	—	350	110	—	390	5	—	—	30	1	5	1190
Gallinula chloropus	400	20	—	—	30+	45	—	65	—	—	—	30	—	10	601
Fulica atra	250+	35	—	—	30+	90	—	52	—	—	—	7	—	10	474
Haematopus ostralegus	3200	900+	+	105	1800	850	18	1170	150	300	71	600	150	65	9379
Vanellus vanellus	1550	50	—	4	600	600	—	285	4	1	7	245	60	65	3462
Charadrius hiaticula	32	5	—	—	7+	10	—	4	10	1	16	5	—	1	89
Charadrius alexandrinus	7	20	—	—	15+	35	12	18	—	—	1	5	—	—	124
Gallinago gallinago	35	5	—	—	65	30	—	48	—	—	—	—	—	—	185
Scolopax rusticola	45+	5	—	—	14+	1	—	5	—	—	—	—	—	2	70
Numenius arquata	130	25	—	—	210	60	—	47	—	—	—	—	—	—	472
Limosa limosa	430	22	—	—	350	300	—	120	—	—	—	220	6	180	1628

Species															Total
Tringa totanus	490	80	—	10	500	230	—	360	10	—	4	195	200	450	2529
Philomachus pugnax	30	7	—	—	9	40	—	28	—	—	—	10	—	100	224
Recurvirostra avosetta	170+	60	—	—	80	35	10	41	—	—	186	975	95	700	2352
Larus fuscus	24	—	—	—	8400	2	—	76	10	10	—	—	—	—	8522
Larus argentatus	3800	6700	—	70	15000	330	500	3300	1155	3000	—	—	—	—	33856
Larus canus	1300	50	—	2	150	45	—	430	20	36	3	—	—	1	2037
Larus ridibundus	10000	2000	—	6100	2100	200	—	545	6	—	7530	6725	1875	3200	40281
Sterna hirundo	285	50	+	1000	34	20	500	86	500	80	640	330	170	160	3855
Sterna paradisaea	75	2+	—	700	4	5	200	3	80	140	225	50	2	—	1486
Sterna albifrons	40	—	+	—	4	2	55	1	5	20	—	—	—	—	126
Sterna sandvicensis	900	—	—	4650	—	—	270	240	—	2	—	—	—	—	6063

Sources:

Texel: Dijksen (1981), Dijksen & Dijksen (1977).
Vlieland: Tekke (1974), Spaans & Swennen (1968), Swennen & De Bruijn (1980).
Richel: C. Swennen (personal communication).
Griend: van Halewijn (unpublished).
Terschelling: F. Zwart (personal communication).
Ameland: Valk (1976).
Engelsmanplaat: Mes, Schuckard & Smit (1980).
Schiermonnikoog: van Dijk (1974).
Rottumerplaat/Rottumeroog: Weijman (1980).
Noord-Holland: Otter (1977, 1979a,b).
Friesland: Zegers, 1975, P. Zegers (unpublished).
Groningen: Prov. Planol. Dienst, Groningen (unpublished).
Dollard: Braaksma & De Wilde (1973).

Additional data on Eider numbers came from Swennen 1976b. Some recent information on Herring Gull and Sandwich Tern numbers from J. Rooth, unpublished.

indication of the start of an influx of migrating birds, the following picture develops.

Autumn migration starts with the arrival of Spotted Redshanks in the second half of June, together with small numbers of several other species of waders. In July many wader species, such as Grey Plover, Knot, Dunlin, Bar-tailed Godwit, Curlew, Redshank, Greenshank and Turnstone, and also Black-headed and Common Gulls arrive in large numbers. Oystercatchers increase sharply in August, whereas dabbling ducks, such as Wigeon and Pintail, arrive from September onwards. In October, Brent and Barnacle Geese are observed in large numbers for the first time. Also an influx of Shelduck is noticeable in that period. The latest species to arrive are Red-breasted Merganser and Goosander, usually in October and November. Since some species arrive when other species are leaving, the total number of birds in the Dutch Wadden Sea is amazingly constant in the period August–February (Fig. 14.5).

For several species, especially waders, the Wadden Sea is an impor-

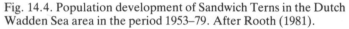

Fig. 14.4. Population development of Sandwich Terns in the Dutch Wadden Sea area in the period 1953–79. After Rooth (1981).

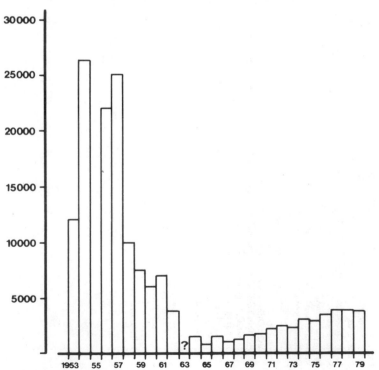

tant moulting area during late summer and autumn. For example, large numbers of Knot and Dunlin moult in the Dutch Wadden Sea (Boere 1976).

Wintering species

In winters without ice considerable numbers of rather few species winter in the Dutch Wadden Sea (Table 14.2). The most abundant birds are Eider, Wigeon, Oystercatcher, Curlew, Knot and Dunlin. Altogether some 600 000–800 000 waders and other coastal birds normally pass the winter in this area. This amoung is similar to the total number of birds wintering together in the German and Danish parts of the Wadden Sea, but the number of birds/km² is higher in the Dutch area. The difference may be ascribed to climatic differences, since ice conditions occur much more frequently in the former areas.

Fig. 14.5. Average number of birds per month in the Dutch Wadden Sea. White columns denote waders, vertically hatched columns grebes, diving ducks, Shelduck, gulls and terns and obliquely hatched ones dabbling ducks and geese. After Smit (1981).

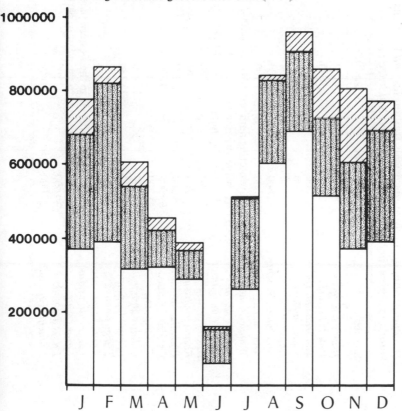

Table 14.2. Number of wintering estuarine birds in the Dutch Wadden Sea as determined by counts from the shore (exception Somateria, Mergus, Larus ridibundus and L. canus). Data on dabbling ducks and Shelduck include numbers present in the Lauwersmeer area. Not listed are species wintering in numbers smaller than 100 individuals

Species	29 Dec 1966	8 Jan 1977	12 Jan 1974	17 Jan 1976	18 Jan 1975
Branta leucopsis	8 200	13 100	17 020	8 480	17 430
Branta bernicla	2 200	6 030	3 800	11 070	6 990
Tadorna tadorna	49 000	44 230	35 180	21 750	30 150
Anas penelope	16 500	30 040	65 540	101 400	76 890
Anas crecca	13 500	3 790	3 380	25 990	13 180
Anas platyrhynchos	17 600	17 600	16 270	26 500	26 370
Anas acuta	5 000	1 100	1 900	3 170	3 370
Somateria mollissima	Approx. wintering number: 104 000–168 000				
Mergus serrator	Approx. wintering number: 12 000				
Mergus merganser	Approx. wintering number: 3 900–9 800				
Haematopus ostralegus	213 000	171 280	181 880	143 550	132 640
Recurvirostra avosetta	200	61	290	480	900
Pluvialis squatarola	1 300	2 300	2 400	2 620	2 110
Calidris canutus	69 000	19 620	69 120	75 500	36 540
Calidris alba	1 200	1 490	1 850	2 260	1 040
Calidris alpina	88 000	71 410	106 140	126 550	109 470
Limosa lapponica	22 000	15 280	18 300	31 160	16 470
Tringa totanus	4 300	8 680	7 730	11 610	9 410
Arenaria interpres	1 300	2 030	1 770	1 720	910
Larus ridibundus	Approx. number wintering in the western part: 45 000				
Larus canus	Approx. number wintering in the western part: 6 400–104 000				
Larus argentatus	39 000	17 100	38 330	26 080	22 290
Numenius arquata	27 000	33 000	51 160	65 200	53 840

When ice winters occur in the Dutch Wadden Sea, which happens on average about once every 5–10 years, numbers of some species in this area also decrease considerably. Shortly after the onset of freezing weather, large numbers of ducks and geese may be observed to leave the area in westerly directions. Fairly large numbers of the normally-occurring species stay. When conditions turn very bad, many animals cease feeding and try to survive by spending as little energy as possible (Swennen, Chapter 10 of this volume).

Spring migration

Spring migration of most species is less conspicuous than autumn migration, since smaller numbers of birds occur together at the same time during spring (Fig. 14.5). Breeding species, e.g. Avocet, Ringed and Kentish Plovers, arrive from March onwards. Species not breeding in the Wadden Sea show peak numbers in the period March–May. In May nearly all species breeding at higher latitudes leave the Dutch Wadden Sea almost completely; the last Brent Geese often stay until early June.

Geographical origins

Birds migrating through and wintering in the Dutch Wadden Sea originate from an area extending from northeast Canada through Greenland, Iceland, Scandinavia and the northern part of the USSR well into Siberia. Individuals with an eastern origin, i.e. Scandinavia, Russia or Siberia, probably predominate although exact figures are not available.

The majority of the species visiting the Dutch Wadden Sea originates from breeding areas around the Baltic and in Scandinavia or further eastward in the northern parts of Russia and western and central Siberia. Some wader species, namely Knot, Sanderling, Dunlin, Redshank and Turnstone, also originate from breeding areas in Iceland, Greenland and/or northeast Canada. For the Redshank this is very clear since the Icelandic subspecies is observed to spend the winter in the Wadden Sea; for other species such as the Knot, the evidence is more circumstantial. Black-headed Gulls also originate from central Europe.

Birds breeding in or migrating through the Dutch Wadden Sea spend the winter mainly along the coasts of the North Sea and eastern Atlantic from immediately south and west of the Wadden Sea (Oystercatcher, Shelduck) through areas in southwest Europe and northwest Africa (Avocet, Bar-tailed Godwit) down to southwest and South Africa (Grey Plover, Sanderling, terns).

Food

Except for a few geese and ducks all Wadden Sea species are carnivores. As a source of food the macrobenthic fauna of the intertidal flats and the shallow water areas appears to be especially important.

The macrobenthic species of the tidal flats of the Dutch Wadden Sea most important in terms of biomass are the mussel (*Mytilus edulis*),

Fig. 14.6. Size selection by some bird species feeding on the same prey species, namely the shore crab (*Carcinus maenas*), based upon analysis of pellets collected at the same high-tide roost in the same period. After Zwarts (1981).

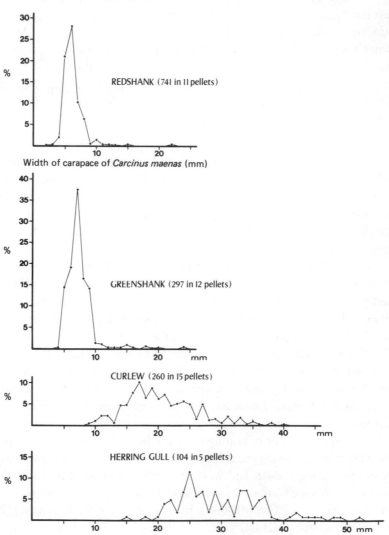

cockle (*Cerastoderma edule*), sand gaper (*Mya arenaria*), Baltic tellin (*Macoma balthica*) and lugworm (*Arenicola marina*). Together they represent over 80% of the benthic biomass (Beukema 1976). The average benthic biomass was estimated as 27 g ash-free dry weight/m^2. The production of the intertidal macrobenthos has been estimated rather conservatively at 35 g ash-free dry weight m^{-2} year^{-1} by Beukema (1981*a*). For the entire area, including the channels and the subtidal shallow water, Beukema estimates the production at about 25 g m^{-2} year^{-1}, including some 5 g of gametes.

Hulscher (1975), Swennen (1976*a*) and Smit (1981) have attempted to quantify bird predation on the benthic fauna. Based on bird counts, a formula relating basal metabolic rate to body weight and a factor relating food uptake to basal metabolic rate, they arrived at an annual consumption by birds of about 4.3, 3.7 and 4.1 g ash-free dry weight/m^2, respectively.

According to K. Rappoldt, M. Kersten & C. Smit (unpublished) these figures have to be increased slightly due to underestimates in bird counts.

At first sight it seems that birds have plenty of food in the Dutch Wadden Sea. However, Beukema (1981*b*) argues that this is not the case. He demonstrates that demersal fish in the Wadden Sea take 3–5 g ash-free dry weight m^{-2} year^{-1} and that benthic invertebrates will take at least 3–4 g m^{-2} year^{-1}, so consumption will amount to at least 10–12 g m^{-2} year^{-1}. Fisheries for mussels and cockles take another 1–2 g. On the other hand, production was estimated at about 25 g m^{-2} year^{-1}, including some 5 g of gametes. This means that only 20 g is produced in a form which can be taken up by birds and other predators. Hence, the difference between production and consumption is 20 g minus 11–14 g which is 6–9 g ash-free dry weight m^{-2} year^{-1}. Since part of the benthic fauna is too deep to be fed upon by birds (e.g. *Mya arenaria*) and another part dies in large quantities during rather short-lasting extreme environmental conditions (e.g. ice winters) this may imply that at least during part of the year, or, in some years, food may be in short supply for birds in the Dutch Wadden Sea.

If food is sometimes in short supply, competition for food will occur. It appears that several mechanisms for reducing or avoiding competition between bird species have been observed to occur in the Dutch Wadden Sea (Zwarts 1981). Spatial segregation is achieved by feeding on different substrates (e.g. Oystercatchers on mussel beds and Bar-tailed Godwits on bare mudflats) and possibly by avoidance of one species by another (e.g. Curlew and Avocet). Temporal segregation is not so much achieved by occurring in different seasons as by feeding at different states of the tide (Swennen 1982). Finally, also segregation by means of different food choice is occurring either by taking different sizes of the same prey species (Fig. 14.6) or by taking different prey species.

Threats to the birds of the Dutch Wadden Sea

The Netherlands' Government has formulated a policy for the Wadden Sea aiming at the preservation and, if necessary, restoration of its natural values. Moreover, about half of the intertidal flats and nearly all of the saltmarshes have been designated as state or private nature reserves, in most cases according to the Nature Conservation Act. This implies that embankments for agricultural purposes are unlikely to occur in future. Embankments for enlargements of harbours or industrial areas stand a somewhat greater chance of being carried out, although in some cases such projects have been cancelled in the past 10 years.

Major problems are formed by disturbance and pollution. Disturbance is caused by all kinds of human activities. These include sailing, walking over the tidal flats to the islands, angling, flying, military training and many others. For most of these activities it is not so much the activity as such which is detrimental, but the large numbers of people taking part in it as well as the multitude of activities which is going on at the same time.

Pollution effects have been demonstrated very convincingly in the years around 1965, when discharge of waste water of a pesticide factory at the Rhine estuary was responsible for widespread bird mortality in the Wadden Sea (Fig. 14.3; Koeman 1971; Swennen 1982). This particular pollution case has been cleaned up, but now the presence of polychlorinated biphenyls (PCBs) is suspected. Koeman *et al* (1973) attributed mortality of Cormorants in other parts of the Netherlands to PCB pollution, and Reijnders (1980) ascribes to PCBs serious effects on reproduction of Harbour seals (*Phoca vitulina*) in the Wadden Sea. Birds in the Wadden Sea also have elevated concentrations of heavy metals (Lida Goede, personal communication) but effects have not yet been demonstrated.

Finally, exploitation of bird populations should be mentioned. Egg collecting is no longer allowed in the Dutch Wadden Sea, while hunting is forbidden on all state property, by far the largest part of the area.

The only activity of importance is the catching of ducks in the last 11 duck decoys in the area. Projectbureau Wadden Sea (1978) estimates that about 30000 ducks are caught annually. Population effects are unlikely, however.

References

Beukema, J. J. (1976). Biomass and species richness of the macrobenthic animals living on tidal flats of the Dutch Wadden Sea. *Neth. J. Sea Res.* **10:** 236–61.

Beukema, J. J. (1981*a*). Quantitative data on the benthos of the Wadden Sea proper. In *Invertebrates of the Wadden Sea*, ed. N. Dankers, H. Kühl & W. J. Wolff, pp. 134–42. Balkema, Rotterdam.

Beukema, J. J. (1981*b*). The role of the larger invertebrates in the Wadden Sea ecosystem. In *Invertebrates of the Wadden Sea*, ed. N. Dankers, H. Kühl & W. J. Wolff, pp. 211–21. Balkema, Rotterdam.

Boere, G. C. (1976). The significance of the Dutch Wadden Sea in the annual life cycle of arctic, subarctic and boreal waders. 1. The function as a moulting area. *Ardea* **64:** 210–91.

Braaksma, S. & de Wilde, W. (1973). De broedvogels van het Dollardgebied. *De Levende Natuur* **76:** 128–32.

Dijk, A. J. van (1974). *De relatie tussen koeien en broedvogels op Schiermonnikoog.* Unpublished report Research Inst. Nature Management, Leersum. 201 pp.

Dijksen, A. (1981). Broedseizoen 1981 gaf Futenexplosie. *Texelse Courant* 13 November 1981.

Dijksen, A. J. & Dijksen, L. J. (1977). *Texel vogeleiland.* Thieme, Zutphen. 246 pp.

Hulscher, J. B. (1975). Het wad, een overvloedig of schaars gedekte tafel voor vogels? In *Symposium Waddenonderzoek. Meded. Werkgroep Waddengebied*, ed. P. A. W. J. de Wilde & J. Haeck, vol. 1, pp. 57–82. Stichting Veth tot Steun aan Waddenonderzoek, Arnhem.

Koeman, J. H. (1971). Het voorkomen en de toxicologische betekenis van enkele chloorkoolwaterstoffen aan de Nederlandse kust in de periode van 1965 tot 1970. Thesis, Utrecht, 136 pp.

Koeman, J. H., van Velzen-Blad, H. C. W., de Vries, R. & Vos, J. F. (1973). Effects of PCB and DDE in Cormorants and evaluation of PCB residues from an experimental study. *J. Reprod. Fert.*, Suppl. **19:** 353–64.

Mes, R., Schuckard, R. & Smit, H. (1980). *Flora en fauna van de Engelsmanplaat.* Meded. 5 Werkgroep Waddengebied, Stichting Veth tot Steun aan Waddenonderzoek, Leiden. 95 pp.

Otter, M. (1977). *Inventarisatierapport Balgzand 1977.* Unpubl. report Nederlandse Vereniging Bescherming Vogels, Zeist. 7 pp.

Otter, M., (1979*a*). *Jaarverslag Balgzand 1978.* Unpubl. report Nederlandse Vereniging Bescherming Vogels, Zeist. 23 pp.

Otter, M. (1979*b*). *Jaarverslag Balgzand 1979.* Unpubl. report Nederlandse Vereniging Bescherming Vogels, Zeist. 28 pp.

Projectbureau Waddenzee (1978). *De interprovinciale struktuurschets voor het Waddenzeegebied.* I. Inventarisatierapport. Leeuwarden. 110 pp.

Reijnders, P. J. H. (1980). Organochlorine and heavy metal residues in harbour seals from the Wadden Sea and their possible effects on reproduction. *Neth. J. Sea Res.* **14:** 30–65.

Rooth, J. (1981). Sandwich tern (*Sterna sandvicensis* Latham). In *Birds of the Wadden Sea*, ed. C. J. Smit & W. J. Wolff, pp. 250–58, Balkema, Rotterdam.

Smit, C. J. (1981). Production of biomass by invertebrates and consumption by birds in the Dutch Wadden Sea area. In *Birds of the Wadden Sea* ed. C. J. Smit & W. J. Wolff, pp. 290–301, Balkema, Rotterdam.

Smit, C. J. & Wolff, W. J. (eds) (1981). *Birds of the Wadden Sea.* Balkema, Rotterdam. 308 pp.

Spaans, A. L. & Swennen, C. (1968). *De vogels van Vlieland.* Wetensch. Meded. Kon. Ned. Natuurhist. Ver. 75. 104 pp.

Swennen, C. (1976*a*). Wadden Seas are rare, hospitable and productive. In *Proceedings of the international conference on the conservation of wetlands and waterfowl*, ed. M. Smart, pp. 184–98. Int. Waterfowl Res. Bureau, Slimbridge.

Swennen, C. (1976*b*). Populatie-structuur en voedsel van de Eidereend
  Somateria m. mollissima in de Nederlandse Waddenzee. *Ardea* **64:** 311–71.
Swennen, C. (1982). De vogels van onze kust. In. *Wadden, duinen, delta*, pp.
  78–100. Pudoc, Wageningen.
Swennen, C. & de Bruijn, L. L. M. (1980). De dichtheid van broedterritoria
  van de Scholekster *Haematopus ostralegus* op Vlieland. *Limosa* **53:** 85–90.
Tekke, M. J. (1974). Ornithologie van Nederland 1972. *Limosa* **47:** 33–50.
Valk, A. (1976). *De broedvogels van Ameland.* Wetensch. Meded. Kon. Ned.
  Natuurhist. Ver. 112. 91 pp.
Weijman, W. (1980). *Verslag bewaking Rottumeroog en -plaat 1978–1979.*
  Report Staatsbosbeheer Groningen. 56 pp.
Zegers, P. M. (1975). Vogels. In *Noord-Friesland Buitendijks. Landelijke
  Vereniging tot Behoud van de Waddenzee*, pp. 36–53. Harlingen.
Zwarts, L. (1981). Habitat selection and competition in wading birds. In *Birds
  of the Wadden Sea*, ed. C. J. Smit & W. J. Wolff, pp. 271–9.
  Balkema, Rotterdam.

15

The Dutch Delta Area

R. J. LEEWIS, H. J. M. BAPTIST &
P. L. MEININGER

Introduction

The Oosterschelde, an estuary in the Delta Area of the southwestern part of the Netherlands with large intertidal areas, will drastically change after 1986.

Part of the tidal area will be dammed up and in the remaining part the tidal amplitude will decrease from 3.40 m to 3.00 m (Fig. 15.1). The estuary is of major importance as a staging, moulting and wintering area

Fig. 15.1. The Dutch Delta Area.

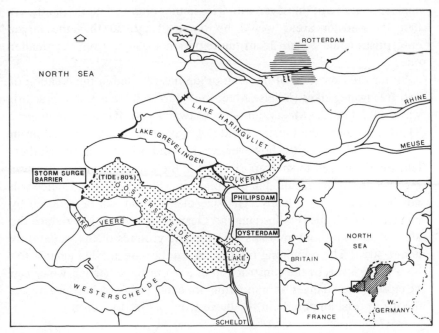

for migratory birds, especially waders. The present as well as the future management policy is determined by the Government's decision to give priority to the natural values of the area. It is important that this policy is given shape in physical planning and management. Therefore, the present functions of the area (ornithological and other) must be known, as well as the character and extent of future changes.

The gathering of this knowledge should not be limited to monitoring developments. Physical planning and management in multi-functional areas like these, where nature, aqua culture and recreation must peacefully co-exist, demand prognoses, in spite of all uncertainties inherent in these.

On the other hand, these uncertainties imply the necessity to exercise physical planning with flexibility, to be able to adapt plans because of increased insight and sometimes changed signals from society.

History

The storm flood of February 1953 caused the inundation of large areas of land. Nearly 2000 people and innumerable cattle were killed. This disaster became the motivation to speed up the already planned reduction of the length of the Dutch coastline. The 'Delta Plan', since 1956 functioning as a governmental law, attended to the enclosure of five of the seven estuaries in the southwest Netherlands. At first the smaller areas would be dammed up, so that the larger enterprises could benefit from the experience gained from the smaller ones.

So far the execution of the Delta Plan has resulted in the creation of two freshwater lakes (Brielse Meer and Haringvliet c.a.), one brackish water lake (Veerse Meer), and one saltwater lake (Grevelingenmeer). The last estuary, the Oosterschelde, was planned to be dammed up in 1978. By the mid-1970s, however, insight and priorities in the Netherlands had changed to such an extent that the execution of the plan was reconsidered. This resulted in new plans in which the Oosterschelde will not be enclosed completely. The size of the basin will be reduced by means of two dams (Philipsdam and Oesterdam), which will reduce the area of the intertidal mudflats (the feeding grounds of the waders) by about 40 %. The remaining tidal basin will be safeguarded against too high water levels by the construction of a storm surge barrier, which can be closed when danger of inundation appears. The open barrier reduces the mean tidal amplitude in the basin from 3.40 m to 3.00 m.

Ornithological research

Regular ornithological research in the Delta Area started in 1963. Monthly counts were carried out on high water roosts along a part of the northwestern coast of the Oosterschelde (Wolff 1973). Those counts were taken as representative for the wader population feeding on the sandflat 'Roggenplaat' in the Oosterschelde.

These counts have been carried out until the present day (R. H. D. Lambeck, unpublished) and will be continued in the future, so as to constitute an important time series for one and the same area.

The first two counts of waterbirds in the whole Delta Area were carried out in 1966–67 (Wolff 1967). Between 1972 and 1976 eight complete counts were carried out (Saeijs & Baptist 1977). From 1976 onwards, the number of counts (especially total counts) have been increased. At first almost complete counts were carried out monthly from September to April and, in addition, since 1979 during the summer months. The most recent counts, especially, give reliable figures for the saline areas – with and without tides (Baptist & Meininger 1984). The large effort involved in these counts meant that less attention could be paid to other research methods. Still, we must point out that fortnightly bird counts from the enclosed Grevelingenmeer are available from 1972 onwards, as well as yearly breeding bird surveys (Beijersbergen & Van den Berg 1980), breeding bird surveys from the Oosterschelde (Baptist & Meininger 1979), yearly surveys of important breeding bird species in the whole Delta Area (P. L. Meininger, unpublished) and many reports on special ornithological aspects. Besides these we have at our disposal a large quantity of as yet unpublished ecological data.

This kind of material has played an important role in the reconsideration of the enclosure of the Oosterschelde and the physical planning and management policy for this estuary (Saeijs & Baptist 1977, 1980).

Developments in the wader populations

When Wolff (1967) carried out his counts in 1966 and 1967 Grevelingen and Haringvliet c.a. were not yet enclosed. Wolff (1967) showed that in August–September 1966 about 152 000 and in January 1967 about 198 000 waders were present in the area, of which in both periods 22 % were in the Grevelingen. After the enclosure of the Grevelingen in 1971, 95 % of the waders disappeared from that area. There are indications that these birds did not leave the Delta Area, but that a rearrangement has taken place over the remaining tidal basins. The results of the total counts confirm this presumption.

Table 15.1 *Numbers of some selected species of waders in the Dutch Delta Area in September 1966–79*

Species	1966	1972	1973	1975	1977	1978	1979
Haematopus ostralegus	55 000	65 000	58 000	62 000	61 000	78 000	92 000
Pluvialis squatarola	2 800	4 200	4 500	4 700	5 500	8 400	6 000
Charadrius hiaticula	3 900	2 300	6 600	3 300	5 000	4 800	3 100
Charadrius alexandrinus	960	980	1 800	700	500	600	640
Arenaria interpres	1 500	1 300	430	800	1 500	1 500	1 600
Numenius arquata	22 000	17 000	11 000	17 000	11 000	22 000	17 000
Limosa lapponica	53 000	5 200	6 900	2 600	5 900	5 300	7 400
Tringa totanus	6 400	9 200	5 100	4 800	3 700	5 600	2 800
Tringa erythropus	1 800	800	1 600	500	2 100	840	1 000
Tringa nebularia	2 000	800	600	800	1 000	800	1 100
Calidris canutus	4 700	4 800	14 000	8 600	11 000	3 700	8 500
Calidris alpina	27 000	42 000	38 000	38 000	45 000	24 000	23 000
Calidris alba	80	100	160	1 700	70	380	400
Philomachus pugnax	2 800	2 400	1 200	500	500	400	150
Recurvirostra avosetta	3 600	1 600	1 800	1 300	1 400	700	900

Table 15.1 shows the total numbers of 15 important species in the Delta Area for September from 1966 till 1979. Most of them appear to fluctuate around a mean level, but some show an increase (Oyster-catcher (*Haematopus ostralegus*), Grey Plover (*Pluvialis squatarola*)), or decrease (Redshank (*Tringa totanus*), Greenshank (*T. nebularia*), and very distinctly, Ruff (*Philomachus pugnax*) and Avocet (*Recurvirostra avosetta*)).

The January counts show a development that agrees only partly with the September situation (see Table 15.2).

The number of Oystercatcher, Grey Plover, Bar-tailed Godwit (*Limosa lapponica*), Redshank and Dunlin (*Calidris alpina*) show an increase. Curlew (*Numenius arquata*) decreases in number.

Of course, organization of counts as well as individual counting skills have improved during the research period. This can, however, only account for a small part of the variation and trends found in the counts.

All enclosures of estuaries in the Delta Area have had profound effects on the distribution of birds over the whole area. Examples have been given by Saeijs & Baptist (1977). At this moment all data available on waders are processed to establish the distribution per season and sub-area for each species. This appears to give some very interesting information. Based on a preliminary evaluation, an example for the Knot (*Calidris canutus*) is given in Fig. 15.2. This shows how Knots use various feeding areas in the Oosterschelde and in the rest of the Delta

Table 15.2. *Development in numbers of some species of waders in the Dutch Delta Area in January 1967–80*

Species	1967	1973	Mean 1975–76	Mean 1979–80
Haematopus ostralegus	81 000	75 000	100 000	110 000
Pluvialis squatarola	1 320	1 500	4 900	6 200
Arenaria interpres	1 750	1 800	1 800	1 760
Numenius arquata	17 200	6 300	8 900	11 500
Limosa lapponica	2 100	6 800	4 600	5 800
Tringa totanus	1 900	2 400	3 200	3 100
Calidris canutus	14 300	18 000	20 000	15 000
Calidris alpina	65 000	45 000	81 000	92 000

Area in the course of the year. This can give clues to possible management measures in the future. This distribution pattern of the Knot cannot be explained yet. Specific research can contribute to the solution of this kind of problem.

Fig. 15.3 shows the development in numbers of Oystercatcher per basin since 1967. The effect of the enclosure of the Grevelingen in 1971 is clearly indicated by the drop in numbers in this basin. In January 1972, the first count after the enclosure of the Grevelingen, numbers of Oystercatcher counted on the Oosterschelde are comparable with those in January 1967, but the numbers on Westerschelde and Krammer–Volkerak appeared to be considerably higher than before. After 1973 the total number of Oystercatchers in the Delta increased gradually, mainly caused by an increase on the Oosterschelde and in spite of a decrease on Krammer–Volkerak and – possibly – Westerschelde. What actually happened with individual birds could have only been revealed by means of marked or ringed birds.

Up till now not many birds have been ringed in the Delta Area. From the few experiments carried out, several interesting facts have become clear already. For instance, it could be shown that nearly all wader species moult in the Oosterschelde (G. J. Slob, personal communication). Moult can be studied in some detail, and should be, because it is very important for management advice to know in what condition birds are at times when certain measures need to be taken. If for some reason a closure of the storm surge barrier is considered, all other consequences, including those for the birds, must be shown. The water level at which this would happen of course determines the seriousness of the consequences for waders. Once the tide has stopped, time is the important factor. How long does it take before a bird has to go and feed elsewhere? In what stage of moulting does this become critical, or impossible? This kind of knowledge is essential for the advisor of the management.

Future work

A study is planned (and in some areas has already started) into the nature of the factors determining the bird numbers and distribution in this area. As the waders are by far the most threatened group the study will concentrate on these birds.

Fig. 15.2. Average numbers of Knot in different parts of the Dutch Delta Area. Key: '——', eastern part of the Oosterschelde; '-----', western part of the Oosterschelde; '----', rest of Delta area; '—·—', total Delta area.

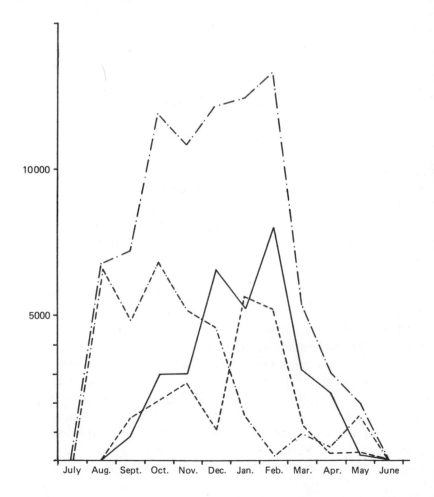

This study can be considered as a kind of environmental impact assessment. There is no time to study the 'black box' between cause and consequence, because the area, as well as the problem, is quite large and answers *must* be known before 1985–86.

The following are questions to be asked:

What is the function of the Delta Area, and the Oosterschelde, as a breeding, wintering, passage and moulting area?
Which ecological factors determine the numbers of birds in this area?
How will waders react on the enclosures?
What will be the influence of the engineering project on total international population sizes and distribution?

Although it may not be possible to answer all questions, a research programme has been set up consisting of four main points:

(*a*) Counts will continue in the same way as they have been organized until now.

The counts comprise the basis of all our bird research. Several institutes and many amateur groups participate in the programme.

Fig. 15.3. Numbers of Oystercatcher in the Dutch Delta Area in January 1967–80. G = Grevelingen; K = Krammer–Volkerak; W = Westerschelde; O = Oosterschelde.

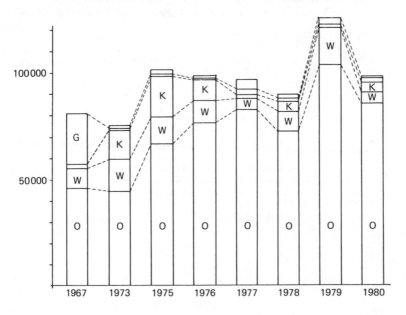

(b) A ringing programme will be started which will provide information on migration, moult, weights and measurements of birds visiting the area. Also it is intended to participate in the colour-marking schemes proposed by M. W. Pienkowski (University of Durham) in 1980. The studies will help to establish spatial and temporal distribution of waders within the Delta area and to reveal relations with other European estuaries.

(c) Studies of several aspects of feeding ecology will be carried out. Firstly, a detailed study of the feeding ecology of some selected species was started in 1979 on a tidal mudflat called 'Slikken van Vianen' (Meire 1980; Meire & Kuyken, Chapter 3 of this volume). It concentrates on density regulation on different types of substrates in relation to distribution of food organisms. Secondly, several other feeding areas in the Oosterschelde are to be studied, but less intensively. In this way we hope to be able to test the results of Meire on the Slikken van Vianen in other circumstances, and find out if extrapolations can be made.

(d) One more factor determines the arrangement of these studies. It is very important for us to have all data available at any moment of the study period. This is so beçause ad hoc advice on all kinds of problems is asked of us. A method of data processing has been worked out by which this demand can be met.

References

Baptist, H. J. M. & Meininger, P. L. (1979). *Broedvogels van het Oosterscheldegebied c.a. in 1978*. Rijkswaterstaat, Deltadienst nota DDMI-79.07.
Baptist, H. J. M. & Meininger, P. L. (1984). *Resultaten vogeltellingen Deltagebied 1975–1980*. Internal report, Rijkswaterstaat.
Beijersbergen, J. & van den Berg, A. (1980). *De Grevelingen. De vogels van een afgedamde zeearm*. Kerckebosch, Zeist.
Meire, P. (1980). Inleidend onderzoek naar de voedseloecologie van enkele steltlopers (Aves – Charadriiformes) op de Slikken van Vianen (Oosterschelde). Thesis, State University of Gent, Belgium.
Saeijs, H. L. F. & Baptist, H. J. M. (1977). Wetland criteria and birds in a changing Delta. *Biol. Cons.* **11**: 251–66.
Saeijs, H. L. F. & Baptist, H. J. M. (1978). Steltlopers in het Deltagebied. *Limosa* **51**: 52–63.
Saeijs, H. L. F. & Baptist, H. J. M. (1980). Coastal engineering and European wintering wetland birds. *Biol. Cons.* **17**: 63–83.
Wolff, W. J. (1967). Watervogeltellingen in het gehele Nederlandse Deltagebied. *Limosa* **40**: 216–25.
Wolff, W. J. (1973). Resultaat van vijf jaar steltlopertellingen op Schouwen. *Limosa* **46**: 21–41.

16

The British Isles

P. R. EVANS

Introduction

The coastline of the British Isles (Fig. 16.1) is long and varied in character. Including the offshore islands it presents some 14 500 km of rocky shore, shingle and sand beaches, and mud-fringed estuaries to coastal birds – but not all are used by waders. Sea cliffs dominate the coastline in certain parts, particularly in north and west Scotland and southwest England. On rocky coasts only the more gently sloping intertidal regions of eastern Britain hold many waders, and then of a more restricted range of species than are found on the softer intertidal sediments. Unlike the coast of much of continental Europe south of the Baltic, that of the British Isles is interrupted by numerous discrete river estuaries. Those of southeast England are particularly muddy, with tidal amplitudes of only a few metres at the mouths of the lowland rivers that flow into them. (Tidal amplitudes are even less along the Netherlands, German and Danish coasts.) The estuaries around the Irish Sea have much greater tidal ranges and consequently greater variation between the area of intertidal land exposed on spring and on neap tides. The largest estuaries in terms of intertidal area, the Wash in eastern England and Morecambe Bay in northwest England, experience lower salinities and greater variations in salinity than most of the Wadden areas of continental Europe, since each of these two British estuaries receives the discharges of four rivers.

In the years 1969–75, counts of waders, gulls and wildfowl were made each month on most of the 133 British and Irish estuaries or bays holding more than 1 km^2 of intertidal land, and on certain stretches of open coast. The results of the 'Birds of Estuaries Enquiry' have been ably summarized by Prater (1981a) in the book *Estuary Birds*, to which readers are referred for detailed accounts of each estuary and bird species.

This chapter sketches first the general importance of British estuaries to waders at different times of year; then provides short reviews of the origins, numbers and periods of movement of the more important

Fig. 16.1. The coastline of the British Isles.

species visiting the British Isles; considers how population turnover might be measured, with reference to detailed studies on the Tees estuary; assesses the importance of climate and food resources in determining the use of British estuaries by shorebirds; and finally outlines some problems requiring further study.

Seasonal changes in numbers of waders in Britain – the general picture

Wader numbers in Britain begin to build up in July and August, with the arrival of non-breeders, failed and finally successful breeders. For the next two months large numbers of a variety of species are present on the Wash in eastern England and on the Ribble estuary and Morecambe Bay in northwest England (Prater 1981a), where many adults moult out of breeding plumage, replacing both their body and flight feathers. Other estuaries are also used as moulting grounds, but for much smaller numbers of birds – thousands rather than tens of thousands. The Ribble may not always have been such an important moulting site for some species, as Holder & Wagstaffe (1930) reported that in the autumns of 1921–29 only *c*. 200 Bar-tailed Godwit were present on the estuary, by comparison with many thousands today.

In late August and September young birds swell the numbers of waders on many British estuaries. Some stay to winter in Britain but others move on quickly, southwards to France and northwest Africa. Total counts of waders often fall in early October, before rising again when influxes of birds arrive from the Wadden areas of Europe. These are chiefly adults that have just finished moulting. At the same time, many birds that have moulted on the Wash and the estuaries of northwest England disperse to the smaller estuaries around the British coasts. It is not clear to what extent waders fly directly from the Wadden Sea to these small estuaries, or whether most move via the Wash (Dugan 1981).

Arrivals on the small estuaries may continue throughout the winter months. In part, these are the result of redistribution of birds within the British Isles, but some movements may originate from the continent of Europe, particularly at the onset of severe weather in December and early January. In January, overall numbers in Britain of several species, e.g. Bar-tailed Godwit and Grey Plover, often rise even in mild winters, possibly the beginning of return movements from further south (Evans 1976; Townshend 1982).

In February and March, adults of many of the species that have wintered in Britain move either southwards from northern Britain (M. W. Pienkowski, personal communication) or directly eastwards, to the Wadden areas. Other species that have spent the winter dispersed along the coasts or even inland (e.g. Curlew) gather into the estuaries at this time. Early April is a time of generally low numbers of shorebirds in Britain, but further immigration occurs in early May, usually of birds moving on quickly to the north and west towards Iceland and Greenland (Ferns 1980a,b; 1981a,b).

The picture thus emerging is one of constant flux in the numbers and origins of birds on a particular British estuary. This makes it extremely difficult to assess the conservation importance of each estuary merely by the maximum count of waders in that site at any one time of year. What is required is a measure of the total number of individuals using it, for however short a period, at some time during the year. This in turn requires estimates of the turnover of populations in different months. Preliminary attempts to assess turnover are described later.

An additional requirement for long-term planning of shorebird conservation, by safeguarding particular estuaries in Britain, is knowledge of the predictability of the movement patterns. Do the same birds move between the same sequences of estuaries in different winters? In particular, are movements similar in cold and mild winters? Although some information on these points has been obtained from recoveries of waders ringed in Britain in the last 20 years, most recoveries of birds carrying metal (numbered) rings have not been in the same non-breeding season as that of ringing. Hence it has been impossible to separate movements within a winter from those between winters. Recently a project has begun to investigate these problems, funded by the Nature Conservancy Council of Great Britain and the European Economic Commission's Environment Programme. The project involves marking of waders with plumage dyes on the moulting areas (in Britain, particularly on the Wash), and using a network of observers along the coasts of the British Isles and other parts of western Europe to record the dates of arrival and duration of stay of marked birds. Also, the proportion of marked birds amongst the population in each estuary at different dates between autumn and spring can be measured. First results are included in the outlines of migration patterns for the commoner shorebird species, summarized below, and have been collated by M. W. and A. E. Pienkowski. Such results are indicated by '(MWP)' in the following section.

The origins and movements of the commoner estuarine waders in Britain

Oystercatcher

The migrations of this species have been reviewed by Dare (1970) and Anderson & Minton (1978) who showed that most of those birds occurring around the Irish Sea coasts in autumn and winter derive from the Scottish, Icelandic and Faeroese breeding populations, whereas those reaching eastern Britain in autumn are chiefly Norwegian breeding birds. Numbers in Britain are highest just before and during the moult period, when more than a quarter of a million are usually present (Prater 1981*a*). After moult, some adults and immatures move away from the Wash, both northwards along the east coast of England and southwards to the Channel coasts (MWP). Since most Oystercatchers remain to winter either in Britain or in the Netherlands, with less than 10% in France (Prater 1981*b*), return migration through Britain is not obvious. Oystercatchers return to their breeding grounds earlier in the year than most waders and desert the estuaries in February and March, except for immature non-breeding birds which may remain through the summer.

Ringed Plover

The British Isles hold more than half of the 30000 Ringed Plover present in mid-winter in western Europe (Prater 1981*b*). These are chiefly British breeding birds and their young, together with immigrants from the Baltic countries and those bordering the North Sea. In August and September and again in May, migrants from the Icelandic and Greenlandic breeding populations pass through, to and from winter quarters in Africa (Taylor 1980). The spring migration is more obvious on the estuaries of northwest England and Wales than elsewhere (Ferns 1980*a*). Some birds passing south in autumn along the English east coast probably return by a more westerly route (Evans 1966; Goodyer & Evans 1980). Some Russian breeding birds pass through eastern England in May (Ferns 1980*a*).

Grey Plover

Those Grey Plover passing through or wintering in Britain have spent the summer in the Siberian arctic, between the White Sea and the Taimyr Peninsula (Branson & Minton 1976). In mid-winter at least 15000 are present, almost a third of the numbers to be found in western Europe. More are present during the autumn moult, particularly on the

Wash and in southeast England (Prater 1981*a,b*). After this, many birds move south or southwest, to be replaced in part in November by birds that have moulted on the Wadden Sea. There is no evidence of northward movements of Grey Plover within Britain, so that those which reach northeast England in autumn have probably flown directly westwards from the coast of Schleswig–Holstein, where colour-marked birds have been seen (MWP). Further arrivals in December in some years are correlated with freezing conditions in the Baltic and along eastern North Sea coasts (Townshend 1982). Numbers in Britain as a whole, but particularly in southeast England, increase in January and February. The wintering birds, together with these late immigrants, leave in March and early April, returning eastwards, at least some to the Schleswig–Holstein parts of the Wadden Sea.

Knot

Although a small proportion of the Knot reaching the British Isles in autumn belongs to the Siberian breeding population, these birds usually move on to winter further south. The bulk of the 250 000 British wintering birds, which comprise about three-quarters of those present in western Europe in winter (Prater 1981*b*), come from northern Greenland and northeastern Canada (Dick *et al.* 1976). Knot arrive in large numbers in August to moult on the Wash and the Ribble estuary, with smaller numbers present on the Dee estuary and Morecambe Bay. The importance of the Ribble as a moulting area has been considerable throughout this century, as records from Wagstaffe (1926) and Oakes (1953) indicate.

After growth of new flight feathers, many birds leave these four estuaries. Most of those from the Ribble probably move to the south of the British Isles, though some may stop at other estuaries around the Irish Sea. Birds from the Wash move not only south but also north along the east coast of Britain, at least as far as the Forth in Scotland. Others cross England to the Irish Sea estuaries in at least some years (Dugan 1981). From November onwards, Knot that have moulted on the Wadden Sea cross to the British Isles. Many arrive at the Wash, but it is not known whether this is a necessary stepping stone to other estuaries, or whether some birds fly to these direct from the Wadden Sea. In spring, birds from the Irish Sea estuaries move northwards to concentrate on Morecambe Bay, particularly in April (Wilson 1973), whilst many of those from eastern British estuaries leave in March for the Wadden Sea, at least some moving south to the Wash before crossing the North Sea.

Sanderling

The British Isles support nearly two-thirds of the 20 000 Sanderling known to winter in western Europe (Prater 1981b). Some of these come from the Siberian breeding population, but a small proportion may originate from Greenland (Prater & Davis 1978). Birds breeding in Greenland also pass through Britain in late summer (July and August) when large numbers are seen in northwest England. Since Sanderling moult chiefly in August and September, most of these birds must move on to moult elsewhere. In eastern England, moulting flocks of many hundred Sanderling are found on the Wash and at Teesmouth. From October onwards, Sanderling disperse from the Wash both northwards to Teesmouth and south and westwards within Britain (MWP). Some individuals continue to move about throughout the winter, but the main departures from wintering areas probably take place in March and April before the influx of migrants from south of the British Isles. These migrants pass once again chiefly through the estuaries of northwest England, particularly Morecambe Bay and the Solway in May (Ferns 1980b).

Dunlin

Numbers of Dunlin in Britain rise during the autumn to a peak in mid-winter and then decline during the spring. In late July and early August, adults that have bred chiefly in Greenland and Iceland move through Britain to moulting and wintering grounds in northwest Africa. Juveniles follow them in August and early September. Adult Dunlin that have bred in northwest Russia arrive chiefly from October onwards, after moulting at the Wadden Sea, but many thousands arrive earlier to moult at the Wash. Juveniles from the Russian population reach Britain in September and October. From October onwards, Dunlin of all ages disperse from the Wash in most directions (MWP). At about the same time, birds begin to arrive from the Wadden Sea, also moving to all parts of the British Isles. By mid-winter, Britain holds about half of the western European wintering population of over one million birds (Prater 1981b). Almost all are of the Russian population (Hardy & Minton 1980). Return movements in late winter and early spring are equally complex. Some Dunlin move south from northern Britain in February and March before heading eastwards to the Wadden Sea (M. W. Pienkowski, personal communication). Others stay on northern estuaries to lay down fat reserves for direct flight in May to Finland and beyond (Goodyer & Evans 1980). Icelandic birds pass through in May calling in chiefly at the northwestern English estuaries and the

Severn. Birds bound for Greenland follow them in late May (Ferns 1981*b*).

Bar-tailed Godwit

The breeding areas from which Bar-tailed Godwits move to winter in Britain comprise northern Scandinavia and western arctic Russia. By late winter about two-thirds of the 100 000 birds in western Europe are found in the British Isles (Prater 1981*b*). Adults arrive to moult in the Wash and the estuaries of northwest England in August. Juveniles arrive in September. After moult, some Godwits leave the Wash, moving both north along the British east coast and south and west within Britain and into France (MWP). Also in late October and November, birds that have moulted on the Wadden Sea move westwards to the British Isles. The largest numbers occur on estuaries around the Irish Sea, on the Wash and in northeast England and southeast Scotland. Influxes on some estuaries occur as late as January, but thereafter numbers fall as birds return to the Wadden Sea to moult into spring plumage (Boere 1976).

Curlew

Almost two-thirds of the 300 000 Curlews wintering in western Europe do so in Great Britain and Ireland (Prater 1981*b*). The coastal wintering populations contain many immigrants, principally from Fenno-Scandinavia. Some British birds move out of the country; others winter inland. Arrivals of adults to moult may take place from July onwards, with juveniles following in September (Bainbridge & Minton 1978). Returns to northern Europe occur principally in March, a month after British breeding birds have moved inland from the coasts.

Redshank

Numbers of Redshank on British estuaries are highest in September, falling to about two-thirds of the Western European total of 150 000 by mid-winter (Prater 1981*b*). Moulting flocks occur on many estuaries in August and September. Besides British breeding birds, these may include Icelandic Redshank which arrive throughout August and September. Juvenile birds marked on the Wash in late summer have moved on quickly, both northwards and westwards; adults have dispersed after the moult, chiefly within the southern half of Britain (MWP). Although flocks on some estuaries, e.g. the Clyde (Furness & Galbraith 1980) apparently remain unchanged in composition throughout a winter, ringing has revealed complex and erratic movements of

Redshank from other areas throughout the winter. Icelandic birds move northwards from March onwards, but no particular concentrations are found on favoured estuaries.

Turnstone
This species is by no means confined to estuaries and coastal soft sediments, though up to 10 000 occur in such habitats in Britain (Prater 1981*a*). The British Isles are close to the southern edge of the wintering range of Turnstone breeding in Greenland and northeast Canada. Adults begin to arrive in late July and stay to moult and winter. Juveniles follow in August. In the same months, Turnstone from Finland pass through the Wash and probably other estuaries in eastern England, pausing only to refuel on their way to Iberia and northwest Africa (Branson, Ponting & Minton 1978). The wintering population is as common on rocky shores as on the estuaries, and forms only a small (but unknown) proportion of those present in the northeast Atlantic. Many winter in Iceland and Norway chiefly on rocky shores.

In spring, Turnstone heading for Greenland and northeast Canada move northwards through Britain in a series of steps in April and early May. Those passing through in May have probably come chiefly from wintering areas south of the British Isles and are travelling to the northernmost parts of the breeding range (Ferns 1981*a*). On Morecambe Bay, the first departures are of birds carrying enough fuel to reach Iceland, whereas birds departing in late May could reach Greenland by a non-stop flight (Clapham 1979).

The estimation of population turnover in an estuary
A first attempt to estimate total numbers of shorebirds using a site during a non-breeding season – and hence to estimate the importance of that site – was made by Bainbridge (in Minton 1975) for the Wash. Using recaptures in the winter of birds that had been ringed there during the moulting period, Bainbridge calculated the proportion, and hence number, of moulting birds of species that had moved away after they had grown new flight feathers. These figures were added to the respective wintering numbers of each species to give minimum total numbers using the estuary during the year. This method assumes that recaptures in winter give reliable indications of the proportion of birds that stayed (which may not be true – see below) and is limited in any case to the few major moulting sites.

On the Tees estuary in northeast England, estimates of population turnover have been made by three methods. The first involved regular

complete counts, usually at least weekly, of several species on the
estuary. By summation of all increases in counts (but disregarding all
decreases) a minimum figure for usage could be obtained. This method
is of no use if rates of immigration and emigration balance. It has been
used with limited success for Knot, Grey Plover and Shelduck in
some years when large 'waves' of immigration or emigration occur-

Fig. 16.2. Counts of Knot, Grey Plover and Shelduck at Teesmouth in
different winters. (Increases are indicated by solid lines, decreases by
dotted lines.) (*a*) Grey Plover, 1976–77; (*b*) Knot, 1978–79;
(*c*) Shelduck.

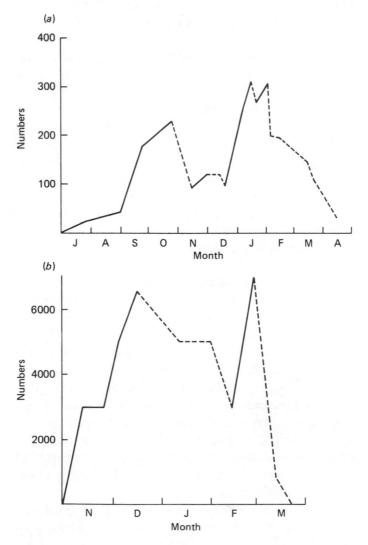

red (Fig. 16.2). In these years, it was obvious that the numbers of individuals using the estuary were several times the peak number recorded.

A second method involved capture–recapture analysis, together with total counts, to estimate net immigration or emigration of birds to/from the estuary. However, it has proved most difficult to determine the proportion of ringed birds in the whole population by conventional trapping techniques. Even if the differing biases of different methods of capture (e.g. use of mist-nets versus cannon-nets, Pienkowski & Dick 1976) are avoided by use of cannon-nets only, there is growing evidence that these do not sample a wader population in a representative way. This arises because the flock is structured, with segregation of age groups, of groups at different stages of moult, and even of subgroups that comprise the same individuals roosting in close proximity on successive occasions (Furness & Galbraith 1980; L. R. Goodyer, personal communication). Only by chance do cannon-nets catch the same part of a roosting flock on successive attempts; on these occasions, very high recapture rates may be obtained – on other occasions few, if any, ringed birds are caught.

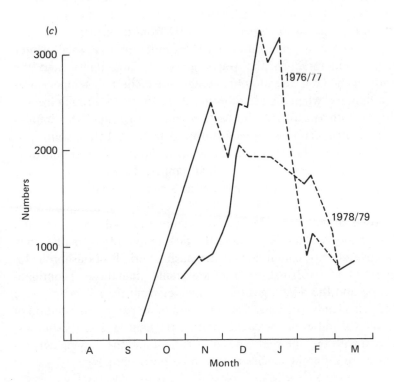

The third and most successful method involved the use of plumage dyes or, better still, individually recognizable combinations of colour rings, to mark birds so that they could be 'recaptured' by direct observation rather than by trapping. By this method, turnover in the Shelduck population on the Tees in the winter of 1978–79 was estimated, even though there was little change in the number of birds present on the estuary between December and February (Fig. 16.2). At least 1800 adults and 600 juveniles moved through in that period, so that at least 4200 individual Shelduck used the estuary that winter, well over double the number present on any one date.

On the larger estuaries, with large wader populations, individual colour-marking is impracticable, because only a small proportion of the population can be so marked and reading of colour-rings or tags at a distance is difficult. However, the use of plumage dyes is practicable and offers much better opportunities for estimating turnover rates than any other method yet available.

Why are the British Isles so important for wintering and passage waders?

The importance of the British Isles for shorebirds arises partly from their geographical position on the migration route from Greenland and Iceland to northwest Africa, and partly from their location on the western seaboard of the continent of Europe, so that their winter climate is moderated by the water mass of the Atlantic and the proximity of the Gulf Stream. On average, the coldest air temperatures occur in January, when the 5° isotherm lies north/south through Ireland and the 3° isotherm follows the eastern coast of Britain. This implies that most British estuaries remain unfrozen for most of the winter, so that waders can feed in the intertidal areas without too much difficulty. Furthermore, the extent of the tidal ranges, between 3 m on the southeast coast and as much as 15 m on the Severn estuary, ensures that, even with onshore winds delaying the ebb of the tide, adequate areas are exposed at low tide. In contrast, large parts of the sand- and mudflats of the Wadden Sea, where the tidal amplitude is smaller, may remain covered throughout 'low' water during periods of onshore gales (Evans 1981). Many British estuaries are smaller than those of continental Europe and the discharges of fresh water from the rivers are more variable. This leads to stronger currents and more patchy distribution of sediments (of different sizes of particles) than in large estuaries. Associated with the different types of sediment and the varying degrees of salinity are different species of intertidal invertebrates.

Most British estuaries provide homes somewhere for the bivalve *Macoma balthica*, often at very high densities in the mud after spatfall, and the mussel *Mytilus edulis*, on firmer substrata. The sand gaper *Mya arenaria* is also widespread, but spat settlement is successful only once every few years, so few of these provide useful food for waders other than Curlew. *Scrobicularia plana* is more restricted in distribution, and is found particularly in the warmer western estuaries in their muddier sections. In contrast, the cockle *Cardium edule* is found at high densities only in the sandiest sediments. Of the other group of molluscs, the gastropods, only the snail *Hydrobia ulvae* is an important food of estuarine waders in Britain.

The most important estuarine worms are the large lugworm *Arenicola marina* and the smaller ragworm *Nereis diversicolor*, which is replaced towards low water mark in more marine conditions by the larger *Nereis virens*. The catworm *Nephthys hombergi* also prefers sandier and more marine conditions than the small *Nereis diversicolor*. In muddier and less saline areas, very high densities of small oligochaete and polychaete worms may be found, for example, on the Tees estuary (Gray 1976). On sandy beaches the polychaetes *Nerine cirratulus* and *Notomastus latericeus* are sometimes common.

Only a few species of crustaceans are important foods for shorebirds in British estuaries: the amphipod *Corophium volutator*, the shrimp *Crangon crangon* and the shore crab *Carcinus maenas*. The last two species move into deeper water in winter, so are taken by shorebirds chiefly in autumn. On sandy beaches, amphipods (*Bathyporeia* spp.) and an isopod (*Eurydice pulchra*) are locally common.

Although most of these invertebrates occur in most British estuaries, their relative abundances vary greatly. However, these alone are not the major determinants of the maximum number of waders present on an estuary at any one moment. Not only must the factors controlling the availability of the prey to the shorebirds be considered, but also the aspects of social interactions between the birds themselves, as has been discussed earlier. It is worth emphasizing again that the same wader species may take a different diet in different estuaries. Few waders are specialists on a certain species of prey; most take whatever common prey species are accessible to their specialized method of foraging (Pienkowski 1981).

Although climatic constraints broadly influence both the distribution and movements of waders within the British Isles, the reasons for concentrations of particular species on particular estuaries at particular times of year are still poorly understood. So too are the roles as refuelling areas of the estuaries which do not hold spectacular numbers at any one time. We are, as yet, some way from a full appraisal of the conservation importance of all British estuaries for shorebirds.

References

Anderson, K. R. & Minton, C. D. T. (1978). Origins and movements of Oystercatchers on the Wash. *Br. Birds* **71:** 439–47.

Bainbridge, I. P. & Minton, C. D. T. (1978). The migration and mortality of the Curlew in Britain and Ireland. *Bird Study* **25:** 39–50.

Boere, G. C. (1976). The significance of the Dutch Wadden Sea in the annual life cycle of arctic, subarctic and boreal waders. Part 1. *Ardea* **64:** 210–91.

Branson, N. J. B. & Minton, C. D. T. (1976). Moult, measurements and migrations of the Grey Plover. *Bird Study* **23:** 257–66.

Branson, N. J. B., Ponting, E. D. & Minton, C. D. T. (1978). Turnstone migrations in Britain and Europe. *Bird Study* **25:** 181–7.

Clapham, C. J. (1979). The Turnstone populations of Morecambe Bay. *Ringing & Migration* **2:** 144–50.

Dare, P. (1970). The movements of Oystercatchers visiting or breeding in the British Isles. In Fishery Investigations, Series II, vol. 25, pp. 1–137. H.M.S.O., London.

Dick, W. J. A., Pienkowski, M. W., Waltner, M. & Minton, C. D. T. (1976). Distribution and geographical origins of Knot wintering in Europe and Africa. *Ardea* **64:** 22–47.

Dugan, P. J. (1981). Seasonal movements of shorebirds in relation to spacing behaviour and prey availability. Unpublished PhD thesis, University of Durham, Durham DH1 3LE, England.

Evans, P. R. (1966). Wader migration in north-east England. *Trans. nat. Hist. Soc. Northumb.* **16:** 126–51.

Evans, P. R. (1976). Energy balance and optimal foraging strategies in shorebirds: some implications for their distribution and movements in the non-breeding season. *Ardea* **64:** 117–39.

Evans, P. R. (1981). Migration and dispersal of shorebirds as a survival strategy. In *Feeding and survival strategies of estuarine organisms*, ed. N. V. Jones & W. J. Wolff, pp. 275–90. Plenum Press, New York.

Ferns, P. (1980*a,b*; 1981*a,b*). The spring migration of waders through Britain in 1979. *Wader Study Group Bull.* **29,** 10–13; **30:** 22–5; **31:** 36–40; **32:** 14–19.

Furness, R. W. & Galbraith, H. (1980). Numbers, passage and local movements of Redshanks on the Clyde estuary as shown by dye-marking. *Wader Study Group Bull.* **29:** 19–23.

Goodyer, L. R. & Evans, P. R. (1980). Movements of shorebirds into and through the Tees estuary. *Co. Cleveland Bird Rep.* **6:** 45–52.

Gray, J. S. (1976). The fauna of the polluted river Tees estuary. *Est. coast. mar. Sci.* **4:** 653–76.

Hardy, A. R. & Minton, C. D. T. (1980). Dunlin migration in Britain and Ireland. *Bird Study* **27:** 81–92.

Holder, F. W. & Wagstaffe, R. (1930). The migrations of Bar-tailed Godwit as observed on the south Lancashire coast. *Br. Birds* **23:** 318–23.

Minton, C. D. T. (1975). *The Waders of the Wash – ringing and biometric studies.* Report for Wash Feasibility Study, Water Resources Board. H.M.S.O., London.

Oakes, C. (1953). *The birds of Lancashire.* Oliver & Boyd, Edinburgh.

Pienkowski, M. W. (1981). Differences in habitat requirements and distribution patterns of plovers and sandpipers as investigated by studies of feeding behaviour. *Verh. orn. Ges. Bayern* **23:** 105–24.

Pienkowski, M. W. & Dick, W. J. A. (1976). Some biases in cannon- and mist-netted samples of waders. *Ringing & Migration* **1:** 105–7.

Prater, A. J. (1981*a*). *Estuary birds.* T. & A. D. Poyser, Calton, Staffordshire.

Prater, A. J. (1981*b*). Wader Research Group Report. *International Waterfowl Research Bureau Bull.* **37**: 74–8.

Prater, A. J. & Davis, M. (1978). Wintering Sanderlings in Britain. *Bird Study* **25**: 33–8.

Taylor, R. (1980). Migration of the Ringed Plover *Charadrius hiaticula*. *Ornis Scand.* **11**: 30–42.

Townshend, D. J. (1982). The Lazarus syndrome in Grey Plovers. *Wader Study Group Bull.* **34**: 11–12.

Wagstaffe, R. (1926). The ducks and wading birds of Southport. *North-West Nat.* **1**: 196–201.

Wilson, J. (1973). Wader populations of Morecambe Bay, Lancashire. *Bird Study* **20**: 9–23.

17

The Atlantic coast of Morocco

M. KERSTEN & C. J. SMIT

Introduction

The Moroccan wetlands are situated between the breeding areas of waders and coastal wetlands of northern and western Europe on the one hand, and on the other hand the important wintering areas in Mauritania, Senegal, Guinea-Bissau and elsewhere along the African west coast. Moroccan coastal wetlands function as a wintering area for many waders and may support even larger numbers during the migration periods. In autumn they function as a moulting area for waders as well. This latter function has been described in detail by Pienkowski *et al.* (1976) and Pienkowski & Knight (1977). This paper deals with the former two functions. Apart from some information on wader numbers in winter, autumn and spring, a modified capture–recapture model will be described, used to determine the number of waders passing through a saltpan and lagoon complex near Sidi Moussa in the spring of 1981. The results of this model allow a preliminary estimate of the number of waders using the Moroccan wetlands.

Moroccan wetlands and their function as staging and wintering areas

Wetlands generally supporting the largest numbers of waders in Morocco are Merja Zerga, Puerto Cansado and the Sidi Moussa–Oualidia area (Fig. 17.1).

Merja Zerga is a large shallow lagoon with a narrow connection to the ocean and a size of 2200 ha. It has been described in detail by Zwarts (1972), Beaubrun (1976) and Kersten & Peerenboom (1978). Table 17.1 gives the results of wader counts in the area. The most numerous species in winter are Dunlin (*Calidris alpina*), Ringed Plover (*Charadrius hiaticula*), and Black-tailed Godwit (*Limosa limosa*). The number of wintering waders generally varies between 33 000 and 49 000, excluding species usually foraging in the saltmarshes (Lapwing (*Vanellus vanellus*), Golden Plover (*Pluvialis apricaria*), Snipe (*Gallinago gallinago*) and Jack Snipe (*Lymnocryptes minimus*)). In some years, the numbers present in Merja Zerga as well as in other areas may be considerably higher; compare the results of Blondel & Blondel (1964) with those of other observers. The reason for the occasional presence of these large numbers has been the subject of much speculation. Pienkowski (1975) suggests it may have been an effect of the severe winter of 1962–63 in Europe.

Puerto Cansado, in the semi-desert of the Tarfaya province, is a lagoon as well. Its total size amounts to *c*. 6000 ha, half of it, however,

Fig. 17.1. The most important wetland sites for waders along the Atlantic coast of Morocco.

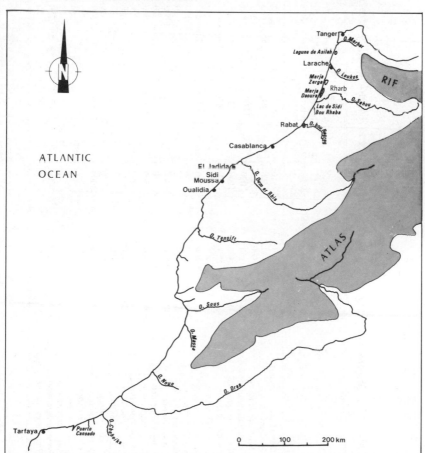

Table 17.1. *Summary of counts of waders in Merja Zerga (Atlantic coast of Morocco)*

Species	22-7-68 Trotignon (1968)	27-9-70 Gooders, in Pienkowski (1972)	12/18-12-70 Zwarts (1972)	9-1-74 Johnson & Biber (1974)	18/22-1-76 Kersten & Peerenboom (1978)	22/23-1-75 Wilson & Hope-Jones (1975)	27/28-1-64 Blondel & Blondel (1964)	1-4-81 Kersten et al. (1981a)
Haematopus ostralegus	150	100	10	72	30		80–100	105
Vanellus vanellus	11	80	2 500	5 000	Thousands	550	40 000–50 000	108
Charadrius hiaticula	20	2 000	20 000	1 000	5 000	1 500	8 000–10 000	472
Charadrius alexandrinus	100	150	2 000		3 000	15	100	3 210
Pluvialis squatarola		400	3 000	1 400	1 600	340	8 000–10 000	
Pluvialis apricaria			800	5 800	Thousands	535	8 000–10 000	
Arenaria interpres		100	30	1	10	5	50	22
Gallinago gallinago		100		10+	Hundreds	75	100	1
Numenius arquata	300	50	2 000	316	180	175	50	112
Numenius phaeopus	1	20	150		10			
Limosa limosa	4 500	5 000	10 000	8 400	9 000	6 000	80 000–120 000	660
Limosa lapponica		300	50	23	20	9	50	4
Tringa totanus	3 500	1 000	3 500	63+	600	140	5 000–6 000	520
Tringa erythropus		250	500	16	30	2	100	5
Tringa nebularia	5	400		5	10	9	100	33
Calidris canutus		100			10		50	20
Calidris minuta		1 000	50		10	200	100	660
Calidris alpina	5	500	8 000	22 000	17 500	21 500	130 000–150 000	15 240
Calidris ferruginea		100						155
Calidris alba		25		4	50	180		8
Philomachus pugnax	200	75		1		380		
Recurvirostra avosetta	200	70	1 900	1 780	1 750	2 470	4 000	2 710
Himantopus himantopus	10	150	3	55	25	10		8

consisting of sebka, an area inundated rarely and totally covered with a crust of crystalline salts. Puerto Cansado has been described in more detail by Blondel & Blondel (1964) and Leah (in Pienkowski 1975). Blondel & Blondel (1964) estimated the wintering wader population in January 1964 at 95 000–115 000. Dick *et al.*, however, arrived at only 5080 in December 1973 (in Pienkowski & Knight 1977) and Johnson & Biber (1974) in January 1974 at 23 000. In September 1972, 25 700 waders were counted (Pienkowski & Knight 1977). Most numerous in the area are Dunlin, Knot (*Calidris canutus*), Grey Plover (*Pluvialis squatarola*), Bar-tailed Godwit (*Limosa lapponica*) and Redshank (*Tringa totanus*) (Table 17.2).

The third important wetland along the Atlantic Moroccan coast is the large saltpan and natural lagoon complex between Sidi Moussa and Oualidia. It is 1 km wide and 40 km long and has been described more fully by Pienkowski (1975), Moser (1981) and the Netherlands Morocco Expedition (NME) (1983). Blondel & Blondel (1964) estimated the wintering wader population at 9000–14 000, Ringed Plover, Grey Plover and Little Stint (*Calidris minuta*) being the most numerous. Pienkowski (1975) estimated the number in September 1972 at 18 000. Kersten *et al.* (1981*a*) counted 6000 waders in the northern part of the area on 1 March 1981. Based on this figure we estimate the total number of waders in the whole area at that time of the year to be between 18 000 and 30 000 birds. Therefore, at the beginning of March, the numbers in this area are probably 1.5–2.0 times greater than the numbers in winter.

Apart from these three most important wetlands several smaller ones can be found along the Atlantic Moroccan coast. Several rivers (oued) run down from the Atlas and Rif mountain chains. Near their mouths, freshwater marshes and saltmarshes can often be found. Further upstream waders locally find foraging possibilities too. Probably the most important river mouth is that of Oued Chebeika, described in detail by Blondel & Blondel (1964), Pienkowski (1975) and Joyes *et al.* (1976), with a wintering population of up to 3000 birds (Blondel & Blondel 1964). In September 1971 the University of East Anglia Expedition counted the same number there; a year later, however, only 400 were present, due to the construction of a dam (Pienkowski & Knight 1974; Joyes *et al.* 1976). Smaller numbers have been counted at the mouth of Oued Loukos near Larache, where marshes as well as a saltpan complex are situated. In some years up to 400 Avocets (*Recurvirostra avosetta*) have been observed to winter in the saltpans (Pineau & Giraud-Audine 1979), which are the most important part of the area for waders. The wintering population at the mouth of Oued Draa can be

Table 17.2. *Estimate of the size and distribution of the wintering wader population along the Atlantic coast of Morocco*

Species	Merja Zerga	Puerto Cansado	Sidi Moussa–Oualidia	Oued Loukos and saltpans near Larache	Oued Draa	Oued Chebeika	Merja Daoura	Oued Sous	Oued Oum el Rbia	Oued Massa	Lagune Aslilah	Oued Sebou	Approximate size of wintering population
Haematopus ostralegus	100	1000	100	5	20	50–450		50	5			20	1 500
Charadrius hiaticula	5000	2000	1400	50	20	200	100	50	10	25	20		9 000
Charadrius alexandrinus	2000	3000	200	100	300	50	5	10	25	50	10		3 000
Pluvialis squatarola	2000	100	800	100	30	up to 175	0–250		10	0–30	10		6 000
Arenaria interpres	20	500	200	10			5	0–30					400
Numenius arquata	200		500	50	5	50	up to 130	0–100		10	0–50	10	1 500
Limosa limosa	10000	2000	500	400			up to 4 500						15 000
Limosa lapponica	50	1000	100	10				10	5	5			3 000
Tringa totanus	3000		1000	300	135	up to 500	up to 370				5	10	7 000
Tringa erythropus	100		50	200	150	up to 1000	up to 100						400
Tringa nebularia	10		50	10									100
Calidris canutus	10	4000	100			50		5					5 000
Calidris minuta	100		500	10		0–20		5	0–100				1 000
Calidris alpina	20000	10000	7000	300	900	500–1000	up to 1000	10	50	15	0–30	10	40 000
Calidris alba	50	500	250	10	5	50	75	20	10	20	10	10	1 500
Philomachus pugnax	50		100	0–25			up to 500	100					500
Recurvirostra avosetta	2000		500	100–400				5			0–150	0–20	3 000
Himantopus himantopus	25		600	150			30						800
Approximate number of wintering waders	45000	24000	14000	1 800–2 100	1 500	1000–3000	500–5 000	250–400	100–200	150	50–250	50–80 000	100 000

Extremely high numbers given by Blondel & Blondel (1964) have been omitted. Wetlands supporting considerably less than 100 wintering waders and inland wetlands have not been inserted. It should be noted that in some years large numbers winter in the plain of the Rharb as a result of which numbers of Black-tailed Godwit in particular are much higher.

Sources on which data from the table were based were: Blondel & Blondel (1964); P. Vandernbulcke (unpublished); Wilson & Hope-Jones (1975); Hovette & Kowalski (1972); Johnson & Biber (1974); M. Kersten, unpublished; Kersten & Peerenboom (1978); Thévenot (1976); M. Thévenot, unpublished; Zwarts (1972). Unpublished information was supplied by IWRB headquarters and A. J. Prater (unpublished).

estimated at 1500 birds (Johnson & Biber 1974). Oued Massa and Oued Sous support 600–800 waders in winter (Blondel & Blondel 1964), and in autumn, numbers are of the same order of magnitude (Pienkowski & Knight 1977). In autumn, numbers may vary greatly from day to day at these two river mouths, pointing to a continuous arrival and departure of birds.

Lakes, situated throughout the country, are generally of limited importance for waders, with the exception of Sidi Bou Rhaba where in autumn and spring hundreds of waders may occur (Thévenot 1976).

After periods of rainfall large areas may become inundated. This happens generally in winter, the amount of rainfall determining the degree of inundation. One of these areas is Merja Daoura, situated along the coast south of Merja Zerga. In other areas in the plain of the Rharb the same thing happens. Blondel & Blondel (1964) counted 100000–150000 waders in this region, mainly Black-tailed Godwits, Lapwings and Golden Plovers.

The northern part of the Atlantic Moroccan coast consists largely of sand beaches, cliffs, rocks and some river mouths. South of Oualidia the coast consists of cliffs, at some places interrupted by sandy beaches. Beaches and rocks offer feeding possibilities to Sanderling (*Calidris alba*), Turnstone (*Arenaria interpres*) and Whimbrel (*Numenius phaeopus*). On the rocks particularly, other wader species can be found as well, especially Dunlin, Ringed Plover, Grey Plover and Redshank. For instance, in March 1981 at somewhat over 10 km of rocky coast south of El Jadida, 678 waders were counted, including 285 Dunlins (NME 1983). The number of waders on beaches and rocky coast is difficult to determine. Because of the relatively low densities on beaches and cliff coasts the total number is probably far below that present in estuaries, lagoons, saltpans and marshes. The wintering population of waders along the Atlantic Moroccan coast, excluding the birds present on rocks and sandy beaches, amounts to about 100000 birds (Table 17.2), a figure coming close to the 139000 waders Prater (1976) estimated to winter along the Atlantic Moroccan coast. These numbers stress the importance of the Moroccan coast as a wintering area, though they are much smaller than numbers wintering in Mauritania (Piersma *et al.* 1980). Pienkowski & Knight (1974) estimated the number of waders south of Merja Zerga in August–September as being 46000. Gooders (in Pienkowski 1972) counted in September 1970 at least 13000 waders in Merja Zerga. Therefore, we estimate the total wader population in autumn to be at least 59000 birds. In the spring there is too little information to give a reasonable estimate of the total number of waders along the Atlantic coast of Morocco. In order to get information on the numbers actually using wetlands along the Atlantic coast of Morocco in March 1981 an expedition was organized whose activities are described below.

Spring migration along the Atlantic coast

Methods

In March 1981 we tried to estimate the number of waders that migrated through a small area of intertidal flats and artificial saltpans along the Atlantic coast of Morocco. The study area forms the northern part of the Sidi Moussa–Oualidia saltpan complex and is situated close to the village of Sidi Moussa (32° 50′ N, 8° 46′ W).

During low tide nearly all birds foraged on the intertidal flats whereas the great majority of the birds spent the high water period resting in the salines. We counted the numbers of waders in the study area at high tide on 1 March, 9 March, 17 March, 21 March and 26 March. During the periods between the counts we captured over 600 birds in the salines with mist-nets. These birds were marked with a dye on the underparts of the plumage and a coloured tape around the aluminium ring. Since we changed from one colour combination of dye and tape to another on the counting dates, every field observation of a marked bird could be traced back to a rather short time interval between two counts.

The proportion of marked birds was sampled on the day of every count and the day after. Most data were collected during low tide when the birds were feeding on the mudflats. With 15–60 × 60 mm telescopes we observed as many birds as possible to check whether they were marked or not. Sample size, number of marked birds and details of the marks were noted. From these data we calculated (1) the number of marked birds still present in the study area at the time of the count, (2) the number of birds emigrating and (3) the number of birds arriving. Emigration and immigration are here defined as the number of birds leaving or entering the study area between two consecutive counts. The equations used to calculate these numbers are given below.

Estimation of the number of marked birds in the study area at any counting day:

$$\hat{M} = N(M/S). \tag{1}$$

Numbers leaving between two consecutive counts:

$$\text{Emigration} = N_1\left(1 - \frac{\hat{M}_2}{\hat{M}_1 + R_2 - R_1}\right). \tag{2}$$

Numbers arriving between two consecutive counts:

$$\text{Immigration} = N_2 - N_1 + \text{Emigration}.$$

Symbols used:

S	Sample size (number of birds checked for marks).
M	Number of marked birds in the sample S.
$\hat{M}_1, \hat{M}_2 \ldots$ etc.	Estimated number of marked birds in the study area at the time of count 1, 2 . . . etc.
$N_1, N_2 \ldots$ etc.	Number of birds counted in the study area during count 1, 2 . . . etc.
$R_1, R_2 \ldots$ etc.	Cumulative number of marked birds released up to count 1, 2 . . . etc.

For the derivations of the equations (1), (2) and (3) the reader is referred to the report of the Netherlands Morocco Expedition (NME 1983). For only three species, Ringed Plover, Redshank, and Dunlin, the numbers of birds captured and marked were sufficient to perform these calculations. The results are discussed below.

Results

The numbers of waders ringed and marked during the periods between the counts are given in Table 17.3. For all three species almost all of the birds were captured between 1 March and 9 March, i.e. between the first and second count. If we assume that these birds form a non-selective independent sample of the population in the study area on 1 March, we are able to calculate which part of that population was still present on the following counting dates. The results are shown in Fig. 17.2, together with the total numbers counted. For all three species it is clear that considerable numbers had left the study area within one week after capture. In Redshank this coincides with an almost identical drop in the number counted, indicating that there was virtually no immigration between 1 March and 9 March. For the other two species, however, the numbers counted on these two dates decreased less than the estimated numbers present since 1 March. The difference on 9 March between the total number and the estimated number present since 1 March represents the number of birds immigrating into the study area between 1 March and 9 March.

For the Ringed Plover the number of birds present since 1 March remained constant at about 250 birds from 9 March to 21 March. The number of Redshanks present since 1 March remained constant at about 290 birds between 9 March and 26 March. According to our calculations all Ringed Plovers that entered the study area after 1 March had left by 21 March, since on that day all the birds counted were thought to be already present on 1 March. Redshanks entering the study area after 9 March had left again by 26 March. These data suggest that the population of these two species consisted of two subpopulations which followed different migratory patterns. Some birds stayed for a rather long time of at least 3 to 4 weeks, but probably much longer since they might have arrived long before 1 March. It is possible that they had

Fig. 17.2. Estimated numbers of birds of three species that stayed in the study area during March 1981 (open circles) and the total numbers present (solid circles) on the same days.

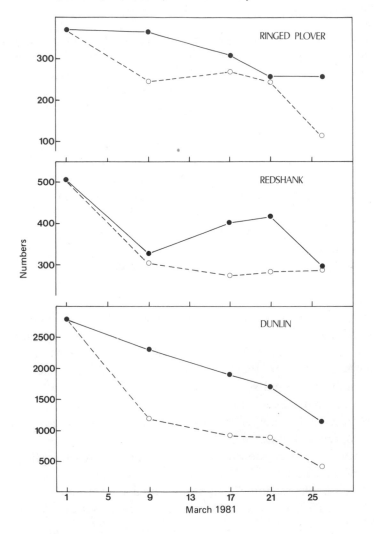

Table 17.3. *Number of waders marked and released in different periods during March 1981. A: 1–9 March, B: 9–12 March, C: 18–19 March, D: 22–26 March*

Species	Period				
	A	B	C	D	Total
Ringed Plover	29	7	0	2	38
Dunlin	287	25	10	30	352
Redshank	46	6	4	19	75

Table 17.4. *Estimates and accuracy (relative standard deviation, RSD) of the numbers of birds immigrating, emigrating and using the study area between 1 and 26 March*

	Ringed Plover		Redshank		Dunlin	
	n	RSD^a(%)	n	RSD^a(%)	n	RSD^a(%)
Numbers on 1 March	372	10	505	10	2821	10
Numbers immigrating	269	59	188	107	1678	64
Numbers emigrating	381	22	405	26	3244	17
Numbers using the study area	641	25	693	30	4499	25

^{a} Assuming 10% RSD in the counting data and no variation in the results of the marked-birds samples. Since we do not know the exact value of the RSD in the counting data, these figures merely indicate how our estimates are affected by random fluctuations in the counts.

spent the winter in this area. The rest of the population consisted of transitory birds that stayed for a much shorter time.

Throughout the observation period birds were continuously leaving and entering the study area (Fig. 17.3). For all three species emigration rates were highest at the beginning and at the end of March. The numbers of birds arriving at and leaving the study area during the observation period are given in Table 17.4. The total number of birds using the study area is calculated by summation of the numbers arriving and the numbers already present on 1 March. The total number using the study area was on average 1.5 times the number present on 1 March.

From the difference between the number of bird-days spent in the study area by all the birds and by the birds present since 1 March we are able to estimate the average number of days that birds arriving after 1 March stayed (Table 17.5). With 12.2 days Dunlins seem to stay for a considerably longer time than Redshanks and Ringed Plovers, which are present for only about one week.

Discussion

The accuracy of the estimated number of birds entering or leaving the study area depends heavily on the reliability of several assumptions which are listed below.

Fig. 17.3. Emigration rates (dark-shaded) and immigration rates (light-shaded) of three wader species in the study area during March 1981.

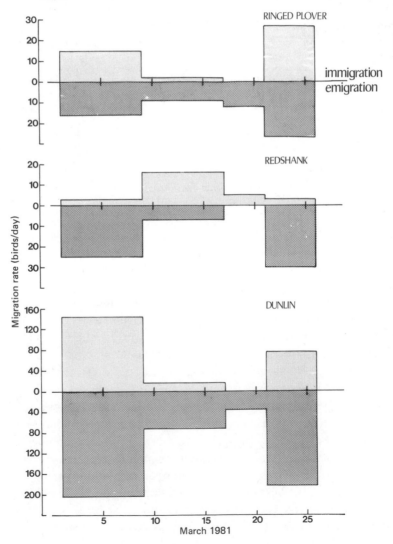

Table 17.5. *Number of bird-days spent in the study area and the average duration of stay of immigrating birds in March 1981*

		Ringed Plover	Redshank	Dunlin
Total number of bird-days spent in the study area	(a)[a]	8 106	9 708	52 377
Number of bird-days for birds already present on 1 March	(b)[a]	6 420	8119	31 873
Number of bird-days for immigrating birds	(c = a − b)	1 686	1 589	20 504
Number of birds immigrating into the study area	(d)	269	188	1 678
Average duration of stay (days) for immigration birds	(e = c/d)	6.3	8.5	12.2

[a] Estimated from the area under the curves in Fig. 17.2.

(1) All the birds had an equal chance of being captured.
(2) The marked birds disperse freely over the whole study area.
(3) Capture and marking did not affect the day of departure.
(4) All birds that arrived after one count did not leave the study area before the next count.
(5) The results of the high-tide counts were accurate.
(6) The determination of the population of marked birds was accurate.

The following comments to these assumptions can be given.

(1) Not all birds had an equal chance of being captured. About 30 % of the birds stayed during high tide in the saltmarshes surrounding the intertidal feeding areas. These were probably the same birds every day and they never entered the catching area. However, since the proportion of the birds roosting in the saltmarshes was rather stable, there is no indication that the migration pattern of these birds differed from that of birds spending the high-water period in the salines.

(2) The marked birds did not disperse freely over the study area either. The proportion of marked birds declined steadily from the northern parts of the intertidal area (close to the catching area) to the southern parts. Most sampling for dyed birds was carried out in the northern part but since the great majority of the individuals of the species mentioned here foraged in these areas, this will not seriously affect the results.

(3) Capture and marking may encourage the birds to leave the study area as soon as possible or on the other hand they may have delayed their departure as a consequence of a negative impact on their condition. Most birds caught had very low weights (NME 1983), indicating a poor condition. This would prohibit them from migrating over any significant distance. Therefore, we looked for marked birds in a small area of artificial saltpans about 3 km to the north of the study area on 4 March and 9 March, but only a few marked birds were observed. We think, therefore, that most marked birds did not leave the study area due to capture. However, birds which died as a consequence of marking would have the same effect on our calculations as birds which left. In fact some marked birds were found dying close to the catching site soon after capture. These were all birds with very low body weights at capture and they were excluded from the analysis. Other birds might have died elsewhere in the area, but we have no indications that this was the case. In Redshank it is very unlikely that this happened since the decrease in the number of '1 March birds' between 1 March and 9 March coincided with an identical drop in the total number (Fig. 17.2). Any significant mortality of marked birds would have caused a positive difference between total number and number of birds present since 1 March.

All birds lost weight, sometimes up to 10 % of the initial weight between capture and release. For the main part this is probably due to faecal production and water loss (Lloyd, Pienkowski & Minton 1979) but the oxidation of energy reserves is almost certainly larger in stressed birds than in unstressed ones. Since the birds use the stored energy reserves during migration, the departure might have been delayed owing to some depletion of the energy reserves. If the birds stayed longer as a consequence of being captured, which is not unlikely, the calculated numbers of emigrating and immigrating birds are underestimated.

(4) Birds entering and leaving between two consecutive counts would result in an underestimation of the numbers migrating through the study area. The time interval between two counts varied from four to eight days. Since immigrating Dunlins stayed on average 12.2 days, this will probably hardly affect the calculations for this species. However, immigrating Redshanks and Ringed Plovers stayed for only about one week (Table 17.5). For these species the calculated numbers emigrating and immigrating during the two eight-day intervals may be seriously underestimated and this implies that the average duration of stay is probably overestimated.

(5) Counting errors affect the results directly since the figures from

the counts are used to calculate emigration and immigration (equations 2 and 3) but also indirectly through the estimated numbers of marked birds in the study area (equation 1). Both random variation and systematic over- or underestimations in the results are involved. Analysis of the effects of counting errors is rather complicated and will be discussed elsewhere (M. Kersten, unpublished). Instead, we will give some results of the error analysis on our data.

Fortunately small systematic errors have little effect. A systematic error of 10 % causes an error of less than 5 % in the estimated numbers emigrating and immigrating over the whole observation period. Numbers counted in a certain area are usually underestimated since the numbers counted in each flock are underestimated and in addition some birds in very small flocks might escape the attention of the observer (Kersten *et al.* 1981*b*). Since our study area is rather small and the birds are easy to find, we think that the systematic error in our counts is less than 10 %.

The effects of random variations in the counting results are much more serious. Experiments in the Dutch Wadden Sea showed that the results of several observers counting the same flock of birds had a relative standard deviation of 20 %, irrespective of flock size and the species involved (Kersten *et al.* 1981*b*). Since the relative standard deviation in the total number of birds in several flocks is smaller because errors in different flocks neutralize each other to some extent, we think that the variation in our counting data is significantly less than 20 %. Nevertheless, Table 17.4 shows that the variation in the estimated numbers emigrating between 1 March and 26 March is still about 20 %. The accuracy of the estimates of the numbers arriving is worse which is mainly due to the fact that these numbers were relatively small.

(6) The variation in our estimates is also affected by random fluctuations in the results of the marked-bird samples. The additional effect of this source of variation is negligible in the case of Dunlin, but in the other two species the variation in the estimated numbers emigrating and immigrating will increase even further. This will be worked quantitatively in a future paper (M. Kersten, unpublished).

From the results obtained in March 1981 we will try to give an indication of the total number of coastal waders that use the areas along the Atlantic coast of Morocco during spring migration. At our arrival on 1 March some 6100 waders were present in the study area. In the species investigated the total numbers migrating through the study area between 1 March and 26 March were estimated to be at least 1.5 times the numbers present initially. Let us assume that this figure holds as an average for the entire period of spring migration for all species and also for the other areas along the coast. If the spring migration period at this latitude lasts about 75 days, this would mean that the total number of waders visiting the Atlantic coast of Morocco is more than 4.5 times the number present at the beginning of March. As we have shown above there are hardly any data on the numbers present at this time of the year. Mid-winter counts revealed that at least 100 000 waders spend winter in Morocco (Prater 1976; this chapter) of which some 14 000 occur in the area between Sidi Moussa and Oualidia. We estimated the total number of waders in this area to be 1.5 to 2.0 times as high on 1 March. This means that the total number of waders visiting the Atlantic coast of Morocco during spring migration must be somewhere in the region of $1.5 \times 4.5 \times 100\,000 = 675\,000$ and $2 \times 4.5 \times 100\,000 = 900\,000$ birds.

At least three million waders spend the winter in West Africa south of the Sahara (Piersma *et al.* 1980; Fournier & Dick 1981). Some birds stay in these areas during summer (Von Westernhagen 1970) while others cross the Sahara inland to the Mediterranean. Therefore, probably two million birds migrate along the Atlantic coast in springtime. Based on this estimate our estimations indicate that some 30–50 % of these birds might use the areas along the Atlantic coast of Morocco as a staging site during spring migration.

The numbers given above are not very accurate of course. They merely give an indication of the order of magnitude of the numbers involved. We hope to collect additional information in April 1982 in order to give more accurate figures for several species separately.

Acknowledgements
During our stay in Morocco we received much assistance from the Administration des Eaux et Forêts. Our gratitude also goes to Michel Thévenot and colleagues at the Institut Scientifique in Rabat. The data presented in this paper could be collected by a joint effort of all members of the Netherlands Morocco Expedition, 1981, i.e. with the aid of the other expedition members Theunis Piersma and Piet Zegers. Their comment, whilst developing the ideas presented here, has been very valuable and stimulating and was highly appreciated. Thanks are due to the staff of IWRB headquarters at Slimbridge and Tony Prater for supplying additional data on wader numbers. The expedition was made possible thanks to financial support from the Beijerink-Popping Fonds, the British Ecological Society, the British Ornithologists' Union, the Landelijke Vereniging tot Behoud van de Waddenzee and the Prins

Bernhard Fonds. Scientific advisors of the expedition were Gerard Boere, Mike Pienkowski and Wim Wolff. Their help and encouragement is greatly acknowledged. Hans van Biezen, Peter Gruys, Thom van Rossum and Wim Wolff read and commented upon an earlier draft of the manuscript. Mrs H. Lochorn-Hulsebos and Mrs B. Soplanit cheerfully faced the drudgery of typing.

References

Beaubrun, P. C. (1976). La lagune de Moulay–Bou-Selham. Etude hydrologique et sedimentologique. *Bull. Inst. scient., Rabat* **1:** 5–37.

Blondel, J. & Blondel, C, (1964). Remarques sur l'hiverage des limicoles et autres oiseaux aquatiques en Maroc. *Alauda* **32:** 250–79.

Fournier, O. & Dick, W. (1981). Preliminary survey of the Archipel des Bijagos, Guinea-Bissau. *Wader Study Group Bull.* **31:** 24–5.

Hovette, C. & Kowalski, H. (1972). *Waterfowl counts in northwest Africa.* Unpublished report to Int. Waterfowl Res. Bureau, Slimbridge, UK.

Johnson, A. R. & Biber, O. (1974). *Dénombrement de la sauvagine hivernant le long de la côte atlantique du Maroc en janvier 1974.* Unpublished report to Int. Waterfowl Res. Bureau, Slimbridge, UK.

Joyes, A., Knight, P. J., Leah, R. T. & Pienkowski, M. W. (1976). The blockage of the Oued Chebeika estuary and its effects on the avifauna. *Bull. Inst. scient., Rabat* **1:** 39–47.

Kersten, M. & Peerenboom, A. M. (1978). Watervogeltellingen in de Merja Zerga, Marokko januari 1976. *Limosa* **51:** 159–64.

Kersten, M., Piersma, T., Smit, C. & Zegers, P. (1981a). Netherlands Morocco Expedition 1981 – some preliminary results. *Wader Study Group Bull.* **32:** 44–5

Kersten, M., Rappoldt, K. & Smit, C. (1981b). Over de nauwkeurigheid van wadvogeltellingen. *Limosa* **54:** 37–46.

Lloyd, C. S., Pienkowski, M. W. & Minton, C. D. T. (1979). Weight loss of Dunlins, *Calidris alpina*, while being kept after capture. *Wader Study Group Bull.* **26:** 14.

Moser, M. E. (ed.) (1981). *Shorebird studies in North West Morocco.* Report of the Durham University 1980 Sidi Moussa Expedition. 100 pp.

Netherlands Morocco Expedition (NME) (1983). *Wader migration along the Atlantic coast of Morocco, March 1981.* Research Inst. Nature Management, Texel and State Forest Service, Utrecht.

Pienkowski, M. W. (ed.) (1972). University of East Anglia expeditions to Morocco 1971 report. University of East Anglia, Norwich, England.

Pienkowski, M. W. (ed.) (1975). *Studies on coastal birds and wetlands in Morocco 1972.* Joint report of the University of East Anglia expedition to Tafaya Province, Morocco 1972, and Cambridge Sidi Moussa expedition 1972. University of East Anglia, Norwich, England. 97 pp

Pienkowski, M. W. & Knight, P. J. (1974). Autumn counts of waders on the Atlantic coast of Morocco. *Bull. Int. Waterfowl Res. Bur.* **38:** 89–92.

Pienkowski, M. W. & Knight, P. J. (1977). La migration post-nuptiale des limicoles sur la côte atltique au Maroc. *Alauda* **45:** 165–90.

Pienkowski, M. W., Knight, P. J., Stanyard, D. J. & Argyle, F. B. (1976). Primary moult of waders on the Atlantic coast of Morocco. *Ibis* **118:** 347–65.

Piersma, T., Engelmoer, M., Altenburg, W. & Mes, R. (1980). A wader expedition to Mauritania. *Wader Study Group Bull.* **29:** 14.

Pineau, J. & Giraud-Audine, M. (1979). *Les oiseaux de la peninsule Tingitane. Trav. Inst. scient., Rabat*, Série Zool. **38.** 132 pp.

Prater, A. J. (1976). The distribution of coastal waders in Europe and North Africa. In *Proceedings of the international conference on the conservation of wetlands and waterfowl*, ed. M. Smart, pp. 255–71. Int. Waterfowl Res. Bureau, Slimbridge, UK.

Thévenot, M. (1976). Les oiseaux de la reserve de Sidi-Bou-Rhaba. *Bull. Inst. scient.*, *Rabat* 1: 67–99.

Trotignon, J. (1968). *Dénombrement de limicoles au Maroc en juillet 1968*. Unpublished report to Int. Waterfowl Res. Bureau, Slimbridge, UK.

Von Westernhagen, W. (1970). Durchzügler und Gäste an der westafrikanischen Küste auf den Inseln der Untiefe Banc d'Arguin. *Vogelwarte* 25: 185–93.

Wilson, J. & Hope-Jones, P. (1975). *Waterfowl counts on the North Atlantic coast of Morocco*. Unpublished report to Int. Waterfowl Res. Bureau, Slimbridge, UK.

Zwarts, L. (1972). Bird counts in Merja Zerga, Morocco, December 1970. *Ardea* 60: 120–3.

18

The Banc d'Arguin (Mauritania)

M. ENGELMOER, T. PIERSMA,
W. ALTENBURG & R. MES

Introduction

Since the late 1960s the Banc d'Arguin in Mauritania has become known as an important wintering area for waders (Von Westernhagen 1968, 1970), with the result that several expeditions have visited the area for ornithological research (Pététin & Trotignon, 1972; Gandrille & Trotignon 1973; Duhautois *et al.* 1974). However, it was not until the autumn of 1973 that the first wader counts and ringing activities were carried out by the Oxford and Cambridge Mauritanian Expedition (OCME) in 1973 (Dick 1975; Knight & Dick 1975). In the winter of 1979–80 a second count followed (Trotignon *et al.* 1980). In the following winter we carried out a third count and tried to investigate the feeding situation for wintering waders.

In this paper the different wader studies carried out on the Banc d'Arguin are summarized and attention is focussed on the role of the Banc d'Arguin in the life cycle of migrating and wintering waders. Suggestions are made as to how a cost–benefit analysis for waders wintering in this area can be performed.

Description of the Area

A national park was founded (the 'Parc National du Banc d'Arguin' – PNBA) in the area of the Banc d'Arguin in July 1979 (Fig. 18.1). The park is situated south of Nouadhibou along the Atlantic coastline between Cap d'Arguin (20° 32′ N) and Cap Timiris (19° 22′ N). The most extensive intertidal flats are found south and west of the fishermen's settlement of Iouik (19° 53′ N, 16° 17′ W). The Banc d'Arguin is only sparsely populated by people. The Imraguen, a fishing people, live on the Banc d'Arguin in only a few (mostly temporary) settlements, each inhabited by 50–100 persons.

Fig. 18.1. The Banc d'Arguin. Left of the coast line is the Atlantic Ocean, to the right the Sahara. Thin lines indicate the boundaries of the mudflats exposed at low tide. The interrupted line borders the 'Parc National du Banc d'Arguin'.

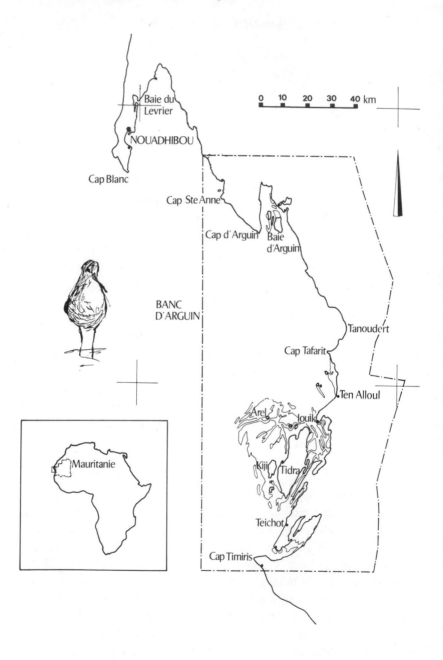

The Banc d'Arguin has an arid (Saharan) climate, although the presence of the Atlantic Ocean and the prevailing northerly winds give it a more oceanic character. Highest average air temperatures are found in September (24 °C) and lowest in February and March (19 °C). The average tidal range lies between 1.20 m and 1.70 m (Dick 1975, and authors' observations).

According to satellite photographs (see NOME 1982) some 540 km^2 of tidal flats are exposed at low tide. Large areas are covered with eelgrass (*Zostera noltii*). On some sandy parts, high in the tidal range, there are patches with cordgrass (*Spartina maritima*) and mangroves (*Avicennia africana*). The tidal flats can be divided into three important substrates, based on descriptions of the areas visited at low tide, the grain size composition of sediment samples and satellite photographs of Landsat Standard Products. (1) The *Muddy Zostera* (390 km^2) is a soft substrate with more than 50 % of the particle weight in grain size classes smaller than 200 μm. In most cases more than 50 % of it is covered with eelgrass. (2) *Sandy Zostera* (50 km^2) is distinguished from the former substrate by having less than 50 % of the particle weight in grain size classes smaller than 200 μm. Eelgrass coverage is mostly less than 50 %, but is always present. (3) The *Barren Sand* is like the Sandy *Zostera*, but without any eelgrass coverage. It is estimated to cover 80 km^2.

The classification accounts for all but 10–20 km^2 of the intertidal area on the Banc d'Arguin. This remnant supports *Arca* fields (large cockle-like bivalves, *Arca senilis*) and *Vaucheria* substrates (an algal species belonging to the Xanthophyllaceae and growing on the mudflats) which are mainly present in the Baie d'Aouatif.

Numbers of wintering waders

By now three fairly complete counts of the wintering numbers of waders on the Banc d'Arguin are available (Knight & Dick 1975; Trotignon *et al.* 1980; Altenburg *et al.* 1983). In the sense that cover of the whole area was attempted, these counts are more or less comparable. In the first count, in rather a short time, most areas were counted from a boat, whereas in the other counts most areas were counted walking along the coastlines with more time available. For the methodological details of the counts the reader is referred to NOME (1982).

In Table 18.1 total numbers of waders on the Banc d'Arguin are presented in relation to the estimated total for the African and European Atlantic and Mediterranean coastlines. The estimates are accounted for in NOME (1982).

Relatively large proportions of the total East Atlantic flyway popula-

Table 18.1. *The mean number of thousands of waders wintering on the Banc d'Arguin (Mauritania) in relation to the estimated totals for the European Atlantic coast, the Mediterranean Basin and the African Atlantic coast*

Species	Mauritania (1) autumn 1973	Mauritania (2) winter 1978/79	Mauritania (3) winter 1980	Total Atlantic coasts Africa (3)	Total Mediterranean Basin (4) (5)	Total Atlantic coasts Europe (4)	Total East Atlantic flyway	Percentage from East Atlantic flyway wintering in Mauritania
Haematopus ostralegus	3.0	6.8	9.2	15.0	1.0	700.0	716.0	1
Himantopus himantopus	—	—	—	10.0	0.5	0.2	10.7	—
Recurvirostra avosetta	—	—	—	20.0	22.0	30.0	72.0	—
Charadrius hiaticula	13.0	136.5	100.0	200.0	2.5	25.0	227.5	55
Charadrius alexandrinus	2.5	6.5	18.0	30.0	23.0	2.0	55.0	18
Pluvialis squatarola	3.5	14.2	24.0	100.0	4.0	50.0	154.0	13
Calidris canutus	130.0	334.0	367.0	400.0	0.1	350.0	750.1	47
Calidris alba	13.0	6.6	34.0	150.0	3.5	25.0	178.5	11
Calidris minuta	5.0	1.0	44.0	150.0	40.0	1.0	191.0	21
Calidris ferruginea	37.0	129.0	174.0	400.0	0.1	—	400.1	39
Calidris maritima	—	—	—	—	—	50.0	50.0	—
Calidris alpina	190.0	705.0	818.0	1000.0	107.6	1500.0	2607.6	31
Limosa lapponica	210.0	538.0	543.0	600.0	0.2	100.0	700.2	77
Numenius phaeopus	3.0	10.5	15.6	50.0	0.1	0.2	50.3	25
Numenius arquata	3.0	10.5	14.2	20.0	7.5	400.0	427.5	3
Tringa erythropus	—	—	—	25.0	1.7	0.5	27.2	—
Tringa totanus	100.0	31.1	70.0	200.0	18.5	150.0	368.5	14
Tringa nebularia	0.5	0.9	1.5	45.0	0.6	1.0	46.6	3
Arenaria interpres	13.0	6.0	17.0	50.0	4.5	25.0	79.5	14
Total/mean	726.5	1936.6	2249.5	3465.0	237.4	3409.9	7113.3	29

(1) Knight & Dick (1975), (2) Trotignon *et al.* (1980), (3) NOME (1982), (4) Prater (1976), (5) Meininger & Mullié (1981).

tion of Ringed Plover (*Charadrius hiaticula*), Kentish Plover (*Charadrius alexandrinus*), Knot (*Calidris canutus*), Curlew Sandpiper (*Calidris ferruginea*), Bar-tailed Godwit (*Limosa lapponica*) and Whimbrel (*Numenius phaeopus*) are found to winter on the Banc d'Arguin.

As far as we are aware at the moment the estimation of wader numbers in the western Palaearctic (covering the western European and northwest African region, and the Mediterranean basin eastwards to Italy and Tunisia), are 3.7 million (Prater 1976) and 5.6 million waders (Scott 1980). However, the delineation of the western Palaearctic region does not fit well for wader migration patterns for two reasons. One is the wintering of Nearctic and mid-Palaearctic breeding birds, as well as western Palaearctic breeding birds in the western Palaearctic region. The other is that western Palaearctic breeding birds also winter outside the western Palaearctic (the African Atlantic coastline south of Mauritania to South Africa). Therefore, the East Atlantic wader flyway (Fig. 18.2)

Fig. 18.2. The East Atlantic flyway for waders.

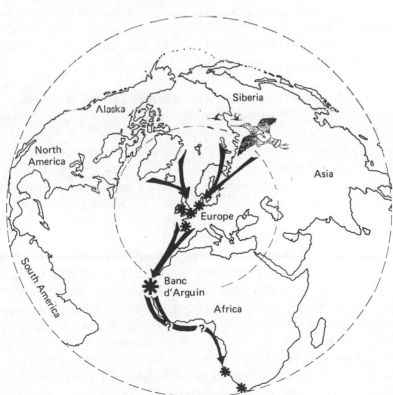

is estimated to contain some seven million waders of which about 30 %
winter on the Banc d'Arguin (Table 18.1). As in the preceding estimates
for the western Palaearctic region, no clear distinctions can be made
here between wader populations migrating along the Atlantic coasts of
Europe, mainly belonging to Nearctic and western Palaearctic breeding
populations, and wader populations crossing the Eurasian continent,
being mainly mid-Palaearctic breeding birds. It can be well imagined
also that different migration routes are used in autumn and spring,
making any distinctions between different migratory populations more
complicated. Another difficulty arises as counting data from only a few
areas along the Atlantic coasts of Africa are available (e.g. Zwarts 1972,
Kersten & Peerenboom 1978 and Kersten *et al.* 1981, Merja Zerga,
Morocco; Blondel & Blondel 1964, Atlantic coastline, Morocco; De
Smet & Van Gompel 1979, Atlantic coastline, Senegal; Fournier &
Dick 1981, Archipel des Bijagos, Guinea-Bissau; Whitelaw *et al.* 1978,
Atlantic coastline, Namibia; Summers 1977 and Underhill, Cooper &
Waltner 1980, Cape Province, South Africa). Of some species men-
tioned in Table 18.1, the numbers wintering in Africa are even less
completely known, because they are also wintering inland in Africa
(Ringed Plover, Little Stint (*Calidris minuta*), Curlew (*Numenius
arquata*), Spotted Redshank (*Tringa erythropus*) and Greenshank
(*Tringa nebularia*) in Moreau 1972, Dowsctt 1980).

The place of the Banc d'Arguin in the wader migration system

Dick (1975) found that the waders on the Banc d'Arguin
originated from both Greenlandic/Canadian and north Euro-
pean/Siberian breeding areas. This was revealed by both biometric data
and ringing recoveries (Dick 1975, 1978). There were two recoveries
further south of waders ringed on the Banc d'Arguin in September 1973:
a Knot in South Africa in June 1975 and a Redshank (*Tringa totanus*) in
Senegal in December 1973. This indicates that some birds present on the
Banc d'Arguin are heading further south. From the trapping data of the
OCME in 1973 (Dick 1975) more indications on migration through
the area in autumn can be extracted. Particularly, juvenile Ringed Plovers
and also Little Stint were observed and trapped more numerously in the
first part of their stay than in October. There was evidence from mean
weight increases that a proportion of both juvenile and adult Knots, part
of the Redshank population, part of the Bar-tailed Godwit population
and a proportion of the adult and juvenile Turnstones (*Arenaria
interpres*) moved on southwards in autumn. In adult Curlew Sandpipers,
moulting was sometimes suspended, indicating southward migration.

There was some evidence from retrapping data that juvenile Sanderlings also moved through. Dick (1975) similarly suspected Greenshank and Whimbrel to migrate further south in autumn. In the Kentish Plover, Grey Plover (*Pluvialis squatarola*) and the Dunlin (*Calidris alpina*) no indications of southward migration in autumn were found. In fact, Fournier & Dick (1981) did not observe any Dunlins in Guinea-Bissau!

In spring, radar observations made by Grimes (1974) in Ghana revealed overland spring migration of waders in northeasterly directions. This suggests trans-Saharan migration routes used in spring by waders that have wintered in west and southwest Africa. It is possible that a proportion of the Mauritanian wintering waders also crosses the Sahara during spring migration. This idea may be supported by the fact that Kersten *et al.* (1981) observed and caught relatively small numbers of Curlew Sandpipers and Little Stints compared to the numbers caught in autumn (Pienkowski 1975; Moser 1981).

Food resources for waders

To quantify the amount of macrobenthic biomass potentially available for waders, accurate samples from all parts and all substrates of the Banc d'Arguin were collected. Samples to two different depths were taken (to 5 cm and to 25 cm). In Table 18.2 the mean biomass per substrate and per sampling depth is given. Between the *Zostera* leaves, prawns and isopods were mainly found but owing to their small size they do not contribute much to the biomass present. In the upper layers of the substrates mainly small polychaetes (*Nereis falsa, Tharyx c. marioni*, Capitellidae) and small bivalves (*Loripes lacteus, Abra tenuis*) occurred. On some places in the Sandy *Zostera* and Barren Sand small specimens of a cockle-like bivalve, *Arca senilis*, were found. The Oystercatcher *Haematopus ostralegus* was the only wader to prey upon these. In the deeper layers of the sand sometimes large polychaetes (*Marphysa sanguinea*) were found on the Sandy *Zostera* and Barren Sand, contributing importantly to the biomass present and eaten mostly by Bar-tailed Godwits.

A mean macrobenthic biomass of about 3.2 g ash-free dry weight/m^2 is found on the Banc d'Arguin as a whole. Compared with other intertidal areas the mean macrobenthic biomass on the Banc d'Arguin is small: 25 g ash-free dry weight/m^2 in the Dutch Wadden Sea (Beukema 1981); *c.* 20 g ash-free dry weight/m^2 in Langebaan Lagoon, South Africa (Puttick 1977); 10–15 g ash-free dry weight/m^2 in Westernport Bay, Victoria, Australia (Robertson 1978). Although the standing crop in February was found to be low, it can be questioned if secondary

Table 18.2. *Biomass of potential food of different substrates for waders on the Banc d'Arguin in February 1980*

Zone of occurrence	Type of animals	Muddy Zostera (390 km²)		Sandy Zostera (50 km²)		Barren Sand (80 km²)		Mean (520 km²)	
		g AFDW/m² (%)		g AFDW/m² (%)		g AFDW/m² (%)		g AFDW/m² (%)	
Eelgrass vegetation	Invertebrates and small fishes	0.14 (7)		0.04 (<1)		— (−)		0.11 (4)	
<5 cm deep in substrate	Macrobenthic species apart from *A. senilis*	1.94 (93)		3.68 (40)		2.32 (46)		2.17 (67)	
<5 cm deep in substrate	Only *A. senilis*	—		0.34 (4)		2.54 (50)		0.42 (13)	
5–25 cm deep in substrate	Large polychaetes	—		5.08 (56)		0.19 (4)		0.52 (16)	
Total biomass		2.08		9.14		5.05		3.22	

For methods and sampling errors see NOME 1982.
AFDW = ash-free dry weight.

production in this subtropical area is high. However, as we argued in NOME (1982), this is not the case and macrobenthic production is also expected to be low.

It is important to clarify the implications of a low macrobenthic biomass on the Banc d'Arguin for the wintering waders. Therefore, we have tried to estimate the amount of biomass eaten by waders from the time the macrobenthos was sampled in February to the time when many waders probably leave the area heading north (end of March, beginning of April), estimated at about 40–60 days. Several assumptions underlie these calculations, which are mentioned in the footnotes to Table 18.3 and are discussed in NOME (1982). A total daily consumption by waders of 25213.7 kg ash-free dry weight for the whole intertidal area or 0.0466 g ash-free dry weight/m^2 is found. In the 40–60 days between the sampling in February and the main departure of waders this would lead to a total macrobenthic consumption of 1.9 to 2.8 g ash-free dry weight/m^2. Assuming no secondary production in the meantime, waders should eat almost everything before departing to the north. The calculations involve normal requirements and do not account for extra intake of waders, needed to gain weight before migration, as they do not include anything other than the macrobenthic food items eaten. This calculation underlines the narrow margins of waders foraging and wintering on the Banc d'Arguin.

Wader foraging

To stress the importance of the macrobenthic food stock for waders on the Banc d'Arguin, data on the diets of waders are presented here. The feeding of waders is restricted to low-water periods, as during high water no suitable feeding areas are available. During low water, information on the prey choice of different wader species was collected and combined with half hourly counts on wader numbers and foraging proportions in sampling areas (0.5–4 ha) during several low-water periods. The food intake of Bar-tailed Godwits was estimated by making observations on success rates and prey sizes (NOME 1982).

As most of the macrobenthic invertebrates present on the mudflats were small-sized, most wader species also took small-sized prey (Table 18.4). However, this was not the case in larger wader species: Oyster-catchers fed almost exclusively on *Arca senilis*; Curlew and Whimbrel on only large crustaceans, fishes (both pipe and flat fishes) and prawns, with Greenshanks eating larger prey than Redshanks. Bar-tailed God-wits ate mostly polychaetes (Table 18.5) with females eating relatively more large polychaetes than males. In relation to the biomass taken

Table 18.3. *Calculation of daily consumption of macrobenthic biomass by waders*

Species	Weight[a] (kg)	BMR[b] (kcal day^{-1} bird^{-1})	Daily intake[c] (g AFDW day^{-1} bird^{-1})	Number	Consumption (kg AFDW day^{-1} species^{-1})
Haematopus ostralegus	0.450	43.96	31.0	7500	232.7
Charadrius hiaticula	0.052	9.23	6.5	100 000	650.0
Charadrius alexandrinus	0.037	7.22	5.1	18 000	91.8
Pluvialis squatarola	0.192	23.75	16.8	24 000	403.2
Calidris canutus	0.126	17.51	12.3	367 000	4514.1
Calidris alba	0.045	8.32	5.9	34 000	200.6
Calidris minuta	0.021	4.79	3.4	44 000	149.6
Calidris ferruginea	0.053	9.36	6.6	174 000	1148.4
Calidris alpina	0.041	7.78	5.5	818 000	4499.0
Limosa lapponica (males)[d]	0.240	27.90	19.7	362 000	7131.4
Limosa lapponica (females)[d]	0.280	31.20	22.0	181 000	3982.0
Numenius phaeopus	0.531	49.55	35.0	15 600	546.0
Numenius arquata	0.800	66.64	47.0	14 200	667.4
Tringa totanus	0.113	16.19	11.4	70 000	797.0
Tringa nebularia	0.250	28.74	20.3	1000	20.3
Arenaria interpres	0.102	15.03	10.6	17 000	180.2
Total				2 247 300	25 213.7

[a] Weights of waders were from Dick & Pienkowski (1979) for adult birds in November.
[b] Basal metabolic rate (BMR) is calculated by the formula for non-passerines of Lasiewski & Dawson (1967): log BMR = log 78.3 + 0.723 log weight (in kg).
[c] Daily intake was taken to be 3 × BMR, divided by 0.85 (digesting efficiency) and 5 (transforming kcal to g AFDW).
[d] The ratio male:female was assumed to be 2:1 (NIOME 1982).

Table 18.4. Diet of waders on the Banc d'Arguin

Wader species	Gastropoda	Bivalvia (not A. senilis)	Arca senilis	Polychaeta (small)	Polychaeta (large)	Crustacea (small)	Crustacea (large)	Fishes
Haematopus ostralegus	++	+	+++	++	++			
Charadrius hiaticula	++	+		++		++		
Charadrius alexandrinus	+	+		++		++		
Pluvialis squatarola	+?	++		+++?	+	+?	++	
Calidris canutus	+?	+?		+		+?		
Calidris alba	+?					+++?		Dead+
Calidris minuta	+?	+?		+++				
Calidris ferruginea	+?	++?		+++		+?		
Calidris alpina	+?	+?		+++		+		
Limosa lapponica (male)		++		++	+			
Limosa lapponica (female)		++						
Numenius phaeopus				++	++	+?	+++	
Numenius arquata					+?	+?	+++	
Tringa totanus	++?					++	+	++
Tringa nebularia						+	++	++
Arenaria interpres	+?	++		++		+++		Dead +

+ + + = very important; + + = important; + = not important; ? = uncertain.

Table 18.5. *Quantitative field observations on the diet of Bar-tailed Godwit (male and female) and Grey Plover on the two substrates of the Banc d'Arguin during daylight low-water periods*

	Bar-tailed Godwit (male)		Bar-tailed Godwit (female)		Grey Plover	
Zostera (Z)/Barren Sand (BS)	Z	BS	Z	BS	Z	BS
No. observations	75	46	52	15	107	21
Small unidentified prey (%)[a]	49	87	46	60	85	90
Large polychaetes (%)	12	11	25	20	4	10
Bivalves (%)	39	2	29	20	8	—
Crabs (%)	—	—	—	—	3	—

[a] Mainly small polychaetes.

large polychaetes were the most important food items; 59 % in males and 80 % in females. Grey Plovers had a broad diet. Pellet analysis and visual observations revealed that several small bivalve species and gastropods but also polychaetes and crab species were eaten (Table 18.5). From these observations it can be concluded that waders were feeding nearly exclusively on macrobenthic prey during daylight.

Waders started foraging as soon as the first mudflats became exposed. Most of them stayed on the mudflats during low water but they did not follow the water's edge. The highest wader densities were measured on Muddy *Zostera* (Table 18.6), mainly owing to the abundance of Knot and Dunlin there.

The total wader densities measured on Sandy *Zostera* and Barren Sand are equal, with highest densities of Knot, Curlew Sandpiper and Dunlin on Sandy *Zostera* and Sanderling, Dunlin and Bar-tailed Godwit on Barren Sand. The mean low-water densities of Knot, Dunlin and Bar-tailed Godwit were the highest ones measured on the Banc d'Arguin.

Although the low-water wader densities are found to be high on the Banc d'Arguin, they are not exceptional. Table 18.6 shows that high low-water densities are also reached in Europe in autumn, where a higher percentage of large waders is involved (Oystercatcher, Bar-tailed Godwit, Curlew and Whimbrel): 15–75 % large waders on the Dutch Wadden Sea compared with 10 % on the Banc d'Arguin (NOME 1982). Feare & High (1977) also reported high wader densities in the Seychelles in winter, mainly consisting of smaller wader species.

On the Banc d'Arguin, high feeding percentages were found (NOME 1982), except for the Oystercatcher on other than *Arca* substrates. In

Table 18.6. *Low-water wader densities (birds/ha) in some intertidal areas in Europe and Africa*

Area	Substrate	Autumn	Winter	Spring	Source
Wadden Sea	Mussel beds	30–80	5–10		L. Zwarts, M. Kersten,
	Sandy areas	5–30	5–10	5–20	& M. Engelmoer (unpublished)
Banc d' Arguin	Muddy *Zostera*		150–200		NOME (1982)
	Sandy *Zostera*		50–75		NOME (1982)
	Barren Sand		50–75		NOME (1982)
Langebaan lagoon	?		15–20		Summers (1977)
Seychelles		50–60	75–100	50–60	Feare & High (1977)
Kenya coast	Sandy areas		1–15		Bryant (1980)

Fig. 18.3 feeding percentages are plotted against the presumed weights of wader species, which were based on the late autumn weights measured by Dick & Pienkowski (1979). As was shown earlier by Pienkowski (1981) in northeast England, foraging activity is smallest in the larger species. The low foraging percentages in large waders on the Banc d'Arguin are mainly the result of a decreasing foraging activity around low water, whereas smaller wader species continue feeding during these periods. Foraging percentages of waders wintering in Europe are also high (Goss-Custard *et al.* 1977; L. Zwarts, unpublished data).

On two substrates on the Banc d'Arguin (Sandy *Zostera*, Barren Sand) a reasonable amount of foraging and prey data for the Bar-tailed Godwit was obtained in order to establish food intake in this species. Because of the remarkable size dimorphism in the sexes of Bar-tailed Godwits, intake was calculated separately for the sexes. By using bill lengths compared with head lengths as a criterion (Smith & Evans 1973) the sexes were identified in the field. For a low-water period males had an intake of 41 kcal on Sandy *Zostera* and 27 kcal on Barren Sand, whereas females had an intake of 65 kcal on Sandy *Zostera* and 129 kcal on Barren Sand. To satisfy their daily needs birds under free-living conditions have to ingest about 2.5–3 times the BMR (Drent, Ebbinge

Fig. 18.3. Mean foraging percentages of waders at low tide on the Banc d'Arguin related to mean wader weights presented by Dick & Pienkowski (1979). Solid circles = Scolopacidae, open triangles = plovers, closed triangle = Oystercatcher.

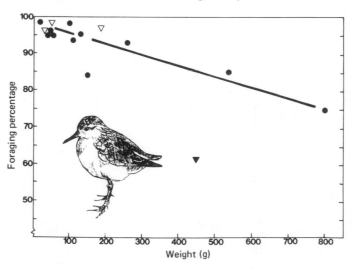

& Weyand 1981). For Bar-tailed Godwits this means a daily need of about 84 kcal in males and 93 kcal in females. It then becomes evident that one low-water period in daylight is not sufficient to approach the daily needs of Bar-tailed Godwits. Even if foraging is extended to the night with an unimpaired intake rate, the daily needs would not be reached in some cases.

All wader species were heard to be present on the feeding grounds during the night and birds followed the tide out when the ebbing tide coincided with dusk.

Conclusions and suggestions

On the Banc d'Arguin, large numbers of wintering waders are present in a relatively small area. As a consequence high wader densities are measured on the feeding grounds. It was observed here that nearly all prey captured by waders consisted of macrobenthic species. However, in February 1980 relatively low densities of small-sized, macrobenthic species were found, leading to low biomasses. The total consumption of macrobenthos by waders on the Banc d'Arguin is estimated to be about 25–75 % which is very high compared with estimates of a total yearly consumption of macrobenthos by birds in the Dutch Wadden Sea: 20–25 % of the macrobenthic biomass available (Smit 1981). Food intake is low during daylight periods, which indicates that waders have to feed at night as well. In fact, waders were present on the feeding grounds at night. Thus, at the moment the high wader numbers seem to be in marked contrast to the relatively small food stock available for waders on the Banc d'Arguin. The nightly foraging situation may be of crucial importance, as it can well be imagined that other prey species become available to waders at night (e.g. more shrimps and flat fishes on the mudflats). Moreover, to develop a proper management strategy for the wader population on the East Atlantic flyway, a clear explanation is needed as to why so many waders winter on the Banc d'Arguin in Mauritania.

Acknowledgements

Gerard Boere and Wim Wolff have been of crucial importance to us, helping during all stages of the expedition. Financially the whole enterprise was made possible by the Prins Bernhard Fonds, Stichting Groninger Universiteitsfonds, Nieuwe Revu, Wereldnatuurfonds Nederland and Stichting Veth tot steun aan Waddenonderzoek. We are greatly indebted to the Government of the Republique Islamique de Mauritanie and the staff of the Parc National du Banc d'Arguin, particularly the Director Mr Abderrahmane Touré, for allowing us to work in the National Park. We thank Marcel Kersten for his home agency and his dedication in ploughing through the early drafts of our paperwork. Thanks go to Jacques and Elizabeth Trotignon and Erik and Veronique Mahe, who have helped us in finding our way in the country. Also thanks go to Michel Binsbergen, William Dick, Rudi Drent, Peter Evans, Mike Pienkowski and others for discussion and criticism. Mrs Poelstra-Hiddinga succeeded in typing the manuscript.

References

Altenburg, W., Engelmoer, M., Mes, R. & Piersma, Th. (1983). Recensement de limicoles et autres oiseaux aquatiques au Banc d'Arguin (Mauritanie). *Gerfaut* 73.

Beukema, J. J. (1981). Quantitative data on the benthos of the Wadden Sea proper. In *Invertebrates of the Wadden Sea*, ed. N. Dankers, H. Kühl & W. J. Wolff, pp. 125–33. Balkema, Rotterdam.

Blondel, J. & Blondel, C. (1964). Remarques sur l'hivernage des limicoles et autres oiseaux aquatiques au Maroc. *Alauda* 32: 250–79.

Bryant, D. M. (1980). Waders on the coast of Kenya: January 1979. *Wader Study Group Bull.* 28: 28–30.

De Smet, K. & van Gompel, J. (1979). Wader counts on the Senegalese coast in winter 1978–79. *Wader Study Group Bull.* 27: 30.

Dick, W. J. A. (ed.) (1975). *Oxford and Cambridge Mauritania Expedition 1973.* Unpublished report, Cambridge, England. 79 pp.

Dick, W. J. A. (1978). Ringing recoveries resulting from the Oxford and Cambridge Mauritanian Expedition 1973. *Wader Study Group Bull.* 24: 31–2.

Dick, W. J. A. & Pienkowski, M. W. (1979). Autumn and early winter weights of waders in north-west Africa. *Ornis Scand.* 10: 117–23.

Dowsett, R. J. (1980). The migration of coastal waders from the Palaearctic across Africa. *Gerfaut* 70: 3–35.

Drent, R., Ebbinge, B. & Weyand, B. (1981). Balancing the energy budgets of arctic-breeding geese throughout the annual cycle: a progress report. *Verh. orn. Ges. Bayern* 23: 239–64.

Duhautois, L., Charmoy, M.-C., Reyal, D. & Trotignon, H. (1974). Seconde prospection post-estivale au Banc d'Arguin (Mauritanie). *Alauda* 27: 241–308.

Feare, C. J. & High, J. (1977). Migrant shorebirds in the Seychelles. *Ibis* 119: 232–39.

Fournier, O. & Dick, W. J. A. (1981). Preliminary survey of the Archipel des Bijagos, Guinea-Bissau. *Wader Study Group Bull.* 31: 24–5.

Gandrille, G. & Trotignon, J. (1973). Prospection post-estivale au Banc d'Arguin (Mauritanie). *Alauda* 42: 313–32.

Goss-Custard, J. D., Jenyon, R. A., Jones, R. E., Newbery, P. E. & Williams, R. Le B. (1977). The ecology of the Wash. II. Seasonal variation in the feeding conditions of wading birds (Charadrii). *J. appl. Ecol.* 14: 701–19.

Grimes, L. G. (1974). Radar tracks of Palearctic waders departing from the coast of Ghana in spring. *Ibis* 116: 165–71.

Kersten, M. & Peerenboom, A. M. (1978). Vogeltellingen in de Merja Zerga, Marokko, januari 1976. *Limosa* 51: 159–64.

Kersten, M., Piersma, Th., Smit, C. & Zegers, P. (1981). Netherlands Morocco Expedition 1981 – some preliminary results. *Wader Study Group Bull.* 32: 44–5.

Knight, P. J. & Dick, W. J. A. (1975). Recensements des limicoles au Banc d'Arguin (Mauritanie). *Alauda* 43: 363–85.

Lasiewski, R. C. & Dawson, W. R. (1967). A reexamination of the relation between standard metabolic rate and body weight in birds. *Condor* **69**: 13–23.
Meininger, P. L. & Mullié, W. C. (1981). Egyptian wetlands as threatened wintering areas for waterbirds. *Sandgrouse* **3**: 62–77.
Moreau, R. E. (1972). *The Palearctic–African bird migration systems.* Academic Press, London and New York.
Moser, M. E. (1981). *Shorebird studies in North West Morocco.* Report of Durham University 1980 Sidi Moussa Expedition. 100 pp.
NOME, 1982. *Wintering waders on the Banc d'Arguin–Mauritania.* Report of the Netherlands Ornithological Mauritanian Expedition 1980. Communication No. 6 Wadden Sea Working Group, Stichting Veth tot steun aan Waddenonderzoek, Texel. 284 pp.
Pététin, M. & Trotignon, J. (1972). Prospection hivernale au Banc d'Arguin (Mauritanie). *Alauda* **40**: 195–213.
Pienkowski, M. W. (ed.) (1975). *Studies on coastal birds and wetlands in Morocco 1972.* Joint report of the University of East Anglia expedition to Tarfaya Province, Morocco 1972, and the Cambridge Sidi Moussa expedition 1972. University of East Anglia, Norwich, England. 97 pp.
Pienkowski, M. W. (1981). Differences in habitat requirements and distribution patterns of plovers and sandpipers as investigated by studies of feeding behaviour. *Verh. orn. Ges. Bayern* **23**: 105–24.
Prater, A. J. (1976). The distribution of coastal waders in Europe and North Africa. In *Proceedings of the international conference on the conservation of wetlands and waterfowl*, ed. M. Smart, pp. 255–71. Int. Waterfowl Res. Bureau, Slimbridge, UK.
Puttick, G. M. (1977). Spatial and temporal variations in intertidal animal distribution at Langebaan Lagoon, South Africa. *Trans. roy. Soc. S. Afr.* **42**: 403–40.
Robertson, A. I. (1978). Trophic interactions among the macrofauna of an eelgrass community. Unpublished PhD thesis, University of Melbourne, Victoria, Australia.
Scott, D. A. (1980). Biogeographical populations and numerical criteria for selected waterfowl species in the western Palearctic. Conf. on the Convention of Wetlands of International Importance. Cagliari, Italy, 24–29 November 1980. Int. Waterfowl Res. Bureau, Slimbridge, UK.
Smit, C. J. (1981). Production of biomass by invertebrates and consumption by birds in the Dutch Wadden Sea area. In *Birds of the Wadden Sea*, ed. C. J. Smit & W. J. Wolff, pp. 290–301. Balkema, Rotterdam.
Smith, P. C. & Evans, P. R. (1973). Studies of shorebirds at Lindisfarne, Northumberland. 1. Feeding ecology and behaviour of the Bar-tailed Godwit. *Wildfowl* **24**: 135–9.
Summers, R. W. (1977). Distribution, abundance and energy relationships of waders at Langebaan Lagoon. *Trans. roy. Soc. S. Afr.* **42**: 483–95.
Summers, R. W., Cooper, J. & Pringle, J. S. (1977). Distribution and numbers of coastal waders in the SW-Cape, summer 1975–79. *Ostrich* **48**: 85–97.
Trotignon, E., Trotignon, J., Baillou, M., Dejonghe, J.-F., Duhautois, L. & Lecomte, M. (1980). Recensement hivernal des limicoles et autres oiseaux aquatiques sur le Banc d'Arguin (Mauritanie). (Hiver 1978/79). *L'Oiseau et R.F.O.* **50**: 323–43.
Underhill, L. G., Cooper, J. & Waltner, M. (1980). *The status of waders (Charadrii) and other birds in the coastal region of the Southern and Eastern Cape.* Unpublished report of the Western Cape Wader Study Group, Cape Town, South Africa.
Westernhagen, W. von, (1968). Limikolen-Vorkommen an der Westafrikanischen Küste auf der Banc d'Arguin (Mauritanien). *J. Orn.* **109**: 185–205.
Westernhagen, W. von, (1970). Durchzügler und Gäste an der westafrikanischen Küste auf der Inseln der untiefe Banc d'Arguin. *Vogelwarte* **25**: 185–93.

Whitelaw, D. A., Underhill, L. G., Cooper, J. & Clinning, C. F. (1978). Waders (Charadrii) and other birds on the Namib coast: counts and conservation priorities. *Madoqua* **11:** 137–50.
Zwarts, L. (1972). Bird counts in Merja Zerga, Morocco, December 1970. *Ardea* **60:** 120–3.

19

The changing face of European wintering areas

W. G. HALE

Introduction

Whilst some species of waders such as Lapwing (*Vanellus vanellus*) and Snipe (*Gallinago gallinago*) winter inland in Europe, most species move to milder coastal areas, and particularly to estuaries outside the breeding season. The way in which wading birds use coastal areas is affected by changes in the coastal environment. This paper considers both long- and short-term changes in the coastal European wintering areas of wading birds and speculates how these changes might (*a*) have affected waders in the past, and (*b*) affect waders in the future.

For the past 2000 years the sea level around the coasts of Europe has been relatively unchanged, and this has meant that the coastlines have been reasonably stable during this period of time. Prior to this there had been considerable variations in sea level. Early in the Pleistocene the bed of the North Sea was dry, with the sea level *c.* 120 m below that of the present day, but at the maximum height of the Flandrian transgression (some 7000 years before present (BP)) the sea level was about 7.5 m above present-day levels.

The fact that the sea level is at present stable does not mean that the estuarine environment and the geomorphology of estuaries is stable. Erosion and deposition continue to alter the face of all our estuaries so that the wintering areas of our populations of wading birds are constantly changing.

Long-term changes

Whilst some evidence exists from which the plant cover and probable climate of nothern Europe can be reconstructed for different periods during the Pleistocene, the best evidence is for the last (Devensian) glaciation which ended between 11 000 and 12 000 years ago. At the height of this glaciation the ice sheet covered the whole of the north of England and the northern tip of Denmark. During this time most, if not all, wading birds probably bred south of this. However, some small areas of Greenland and Iceland were apparently unglaciated and it is possible that occasionally these may have been colonized by land birds. Despite this it is unlikely that they played any significant part in the evolution of the populations of high-arctic waders.

At the height of the glaciation any Greenland breeders would have had to make a journey of over 3000 miles non-stop, much of it against the southerly winds over the ice cap. Iceland would not have provided suitable staging posts because the ice sheet there would have prevented tides and thus there would have been no intertidal zone to provide food for a migrating population of birds. Again the periods of isolation of such populations would have been such that speciation would have occurred, although it is just possible that these populations did exist separately and were obliterated at the height of the Devensian Ice Age, some 25 000–30 000 years BP. Since the species possibly involved (Knot (*Calidris canutus*), Turnstone (*Arenaria interpres*) and Sanderling (*Calidris alba*)) were also present in Europe, this probably indicates that any Greenland refuges were insignificant in the evolution of these species or in the development of migration routes. It is very doubtful if these populations, if they existed, could have played any part in the evolution of migration routes as suggested by Prater (1981).

Immediately to the south of the Devensian ice sheet, rivers which now drain into the southern North Sea emptied into the Atlantic via an estuarine complex in the region which is now the mouth of the English Channel. Further south rivers followed much the same valleys as at present. Because of the lower sea level the estuaries of these rivers were nearer to the edge of the continental shelf and it would be reasonable to assume that because of their long existence most of them would have been shallow estuaries with extensive mudflats or deltas. The more important of the estuaries discharging into the Atlantic were probably those of the Loire and Garonne in France, the Minho, the Douro, Tagus and Guadiana in Portugal and the Guadalquivir in Spain. Only the Portuguese and Spanish estuaries would be likely to hold wintering waders at the height of the Devensian Glaciation, though in the Mediterranean the Ebro in Spain and Rhône in southern France were probably also important.

River valleys and estuaries buried beneath the ice sheet were scoured out and deepened so that as the ice receded and new estuaries were formed, these lacked the extensive mudflats of mature estuaries, and thus provided only limited feeding areas for waders. With the rise in sea level resulting from the return of melt-water to the sea, the coastline moved inland and nearly all European estuaries took on the nature of drowned valleys. Bar-built estuaries were virtually non-existent and mudflats almost entirely absent. In some, more northerly areas, fjords were formed, but again these had even more limited areas for wader feeding.

At the height of the glaciation the rivers of central Europe flowed into the Caspian and Black Seas, which were linked and much more extensive than at present. These drained into the Mediterranean Sea which was consequently less saline and because of the lower sea level less extensive. The Caspian and Black Seas were probably ice-bound in winter and unsuitable for wading birds but the Mediterranean was probably much better for waders than it is now and supported large wintering populations of waders and wildfowl. Evidence for this comes from fossil collections from the Grotte di Romanelli (Blanc 1921) and the Har Dalam Cavern collection from Malta (Bate 1916). Climatically the Mediterranean at the height of the last glaciation was similar to the western British Isles and southern North Sea today, an area which supports large populations of wintering waders.

Shorter-term changes

The ice ages were responsible for changes in the estuaries long after the ice had receded. Eustatic rise of sea level and isostatic land movements altered the relative sea level over long periods of time and these fluctuations have been documented by Tooley (1974). In the past 9000 years 10 major transgressions have been recorded in the north of England (Fig. 19.1) and these are closely correlated with similar changes recorded elsewhere in Europe. However, it does not take an ice age to alter the configuration of an estuary. As it matures, and silt is washed down the river, sand-banks are formed, mudflats extend and a habitat develops that is suitable for large populations of invertebrates which are accessible to waders between the tides. As the estuary matures further, banks often grow higher, and are drier at low tide. Then they do not have the organic content necessary to support the invertebrates, and the populations of these organisms fall. Waders then move elsewhere for their winter food supplies. Developing saltmarsh may also reduce the available areas of mudflats and this change can be very rapid. On the Ribble estuary, saltmarsh extended seawards a distance of over 800 m in 10 years at the end of the last century. Whilst on the one hand reclamation for industrial purposes removes some wader feeding areas, others are maintained by regular dredging of sea lanes in the major navigable estuaries. There can be no doubt that human activities influence greatly the changes occurring in estuaries in the short term. Apart from reclamation and the possible beneficial effects of dredging, both disturbance and pollution affect waders adversely once the birds reach the estuary. Fluctuating numbers and physical changes occurring in the breeding areas can also affect estuarine numbers, and changes on either breeding or wintering areas, or both, can affect the migration routes.

Changing wader populations

At the height of the Devensian Glaciation the tundra probably occupied a greater area than at any other time and this is shown in reconstruction of the plant cover, e.g. Frenzel (1968). It is likely that at this time the populations of tundra-breeding waders, such as Knot, Sanderling, Turnstone, Dunlin (*Calidris alpina*), Bar-tailed Godwit (*Limosa lapponica*), Golden Plover (*Pluvialis apricaria*) and Ringed Plover (*Charadrius hiaticula*), were larger than at any other period of their history. If this was so, then the existing estuaries of western Europe and North Africa had to accommodate much larger winter numbers than a more extensive estuarine system accommodates now. If this is accepted, it follows that present winter populations are likely to

Fig. 19.1. Relative sea level changes in northwest England from 9200 BP to the present. Continuous line = relative movement of mean high water of spring tides. Broken line = relative movement of mean sea level; LI–LX = marine transgressions. After Tooley (1974).

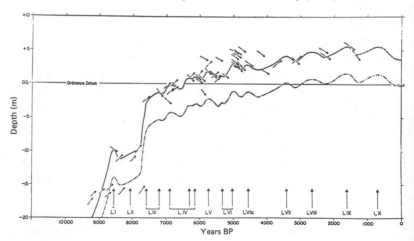

be well below the carrying capacity of the estuaries as a whole. As the ice receded and the climate improved, the tundra area and its wader breeding populations were reduced in size. Almost certainly this happened during interglacials and also interstadials and the tundra area, and the wader populations associated with it, were at a minimum during the height of the Ipswichian Interglacial. At this time tundra refuges were restricted to an area in North Greenland and Ellesmere Island, a second smaller area in the region of Verkhoyansk and two small areas in Alaska (Fig. 19.2) but probably populations of temperate waders were at their highest, because of the extent of the temperate zone.

The extension of the polar ice sheet did not merely move the bands of vegetation types to the south, nor did the amelioration of the climate move them all north again. At the height of the last glaciation loess tundra, a form of cold steppe, occupied a band across Europe north of the Alps. There is no present-day equivalent of this vegetation type (Moreau 1954) but fossil Great Snipe and Common Snipe are recorded from it (Schmidt, Koken & Schuz 1912). Because of the slow spread of trees as the ice receded, grasslands occupied large areas in climates where the climax vegetation was to be deciduous woodland. Therefore the breeding populations of waders would have been much more extensive than would be appreciated by translating climatic data into

Fig. 19.2. Tundra refuges at the heights of the last interglacial (black) and the last glacial (shaded). Colonization of more northerly breeding areas is shown by a thick black line, present-day southerly migration routes by a thin line.

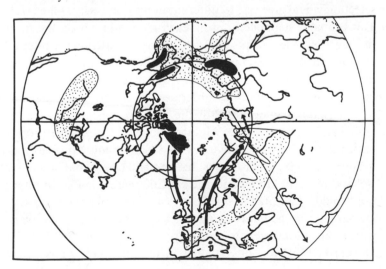

present-day vegetation types and estimating wader populations from these.

In more recent times, the clearing of woodlands in Europe and particularly the deafforestation of the uplands again favoured the expansion of wader populations (Hale 1984) so that overall present-day wader numbers are probably greater than at the height of the most recent climatic optimum in the Atlantic period some 4500–7600 years ago.

Expansion and contraction from the Pleistocene refuges

There can be little doubt that Europe and western Asia were populated by waders which had their origins in the Greenland/ Ellesmere Island interglacial refuge (Fig. 19.2). North America, too, probably was colonized by some species of wader from Greenland/Ellesmere Island, but clearly the Alaskan refuges provided many species. Eastern Asia was probably colonized from the Verkhoyansk refuge. From Greenland, Dunlin, Knot and Bar-tailed Godwit spread first into Europe, then into western Asia, as did the Ringed Plover, Sanderling and Turnstone. The Golden Plover probably contacted the Lesser Golden Plover (*Pluvialis dominica*) extending westwards from the Verkhoyansk refuge, and because of their similar ecology each excluded the other from its established range. Where species met no ecological competition, for example the Dunlin and Ringed Plover, so their expansion continued eastward. Extending ice sheets and isolation of suitable habitat at the height of the last glaciation very probably isolated populations of each of several species, and selection would give rise to different morphs in each refuge (Hale 1980). This probably happened during each glaciation but on the recession of the ice only the morph of each species resident in western Europe retreated to the Greenland/Ellesmere Island interglacial refuge. Other forms were probably either wiped out or isolated in one of the other interglacial refuges so that subsequently, on the next expansion of the ice, speciation had occurred by the time they again came together, e.g. Golden Plover and Lesser Golden Plover. In western Europe, tundra waders, e.g. Knots, Turnstones and Sanderlings, probably retreated north through the British Isles, Iceland and Greenland before becoming isolated in north Greenland and on Ellesmere Island. Another population would be isolated in northern Scandinavia with the recession of the ice but later in the interglacials it is likely that this population would be wiped out as it failed to find a refuge in either Greenland or Verkhoyansk. However, interstadials would often fail to push these wader populations sufficiently far north for this to occur, and this may have been the case in the Atlantic period (4500–7600 years BP), when the climate improved markedly, but probably not so much as to drive the Dunlin and Ringed Plover out of Scandinavia or at least push them no further than the Taimyr Peninsula.

During the Atlantic period tundra forms were probably pushed north into Iceland and Greenland in the west, and in the east into northern Scandinavia or even further east into the USSR. These populations may well have remained separate for up to 7000 years. During at least the first half of this period the British Isles and southern Scandinavia were tree-covered and little wader habitat existed; trees existed throughout this time to a height of between 800 and 1000 m OD so that there would be little open habitat, and that present was probably occupied by the Dotterel *Eudromias morinellus*.

It has been suggested elsewhere (Hale 1980, 1984) that populations of waders that were separated for periods insufficiently long to produce new species gave rise to secondary hybridization on coming together again and produced populations which were morphologically distinct from the two parent populations. Such situations almost certainly arose on numerous occasions throughout the Pleistocene and at least sometimes produced genetically more viable populations. The last occasion when this might have happened could well have been through the deafforestation of the uplands, an event which was largely man made.

Though some peat formation began during the Atlantic period on British uplands, probably no large, open bogs suitable for upland breeding waders such as Dunlin and Golden Plover occurred until much later. The warm and dry Sub-Boreal, an unsuitable climate for sub-arctic waders, lasted until about 2500 years ago, and whilst this was followed by the wetter and cooler Sub-Atlantic, it may well have been the onset of the Little Ice Age in the sixteenth century which resulted in sub-arctic birds breeding for the first time on uplands which, now cleared, grazed and burned regularly, resembled their normal tundra habitat. Here secondary hybridization between the Icelandic/Greenland individuals and birds of Scandinavian origin may well have given rise to 'southern' forms of both Dunlin and Golden Plover (Hale 1984).

There is no reason to believe that at this time the migration routes of the Iceland/Greenland and Scandinavian populations of these species were different from those of the present day. Since populations of both Dunlin and Golden Plover, from Iceland and Scandinavia, pass through the British Isles in autumn and spring, the possibility of association and subsequent interbreeding between the two populations clearly existed.

It has been argued (Hale 1971, 1980, 1984) that in the Redshank (*Tringa totanus*) in particular, and in the Ringed Plover, Golden Plover, Dunlin and Black-tailed Godwit (*Limosa limosa*), hybridization has resulted in the failure of many individuals to complete the normal spring moult of body feathers. This is the main morphological feature of 'southern' populations. In the Redshank, British breeders may vary between a full breeding plumage and a full, unmoulted winter plumage, with no apparent disadvantage to birds having the latter. Ringed Plover and Black-tailed Godwit are nearly as variable, though Golden Plover and Dunlin are less so. It can be assumed that breeding plumage evolved initially because it was of value. Birds lacking a breeding plumage, which breed apparently as successfully as others, probably have other significant advantages which more than compensate for its lack, or are breeding in a situation where breeding plumage is no longer of advantage; the former is the more likely. That the Redshank population of individuals with a reduced breeding plumage is at an advantage to at least one of the presumed parent populations is shown by the spread into Scandinavia of large cinnamon birds with little breeding plumage. Birds with these characters are rapidly replacing small birds with a full breeding plumage in Scandinavia (Hale 1971), and a similar spread of large Ringed Plovers with reduced breeding plumage is also taking place there (Vaisanen 1969).

One character which these spreading populations possess which is apparently of advantage over that of the presumed parent populations is that they breed earlier and have therefore the possibility of relaying more often. Another apparent advantage is that the spreading, hybrid population is much less migratory than the parent population, and in the case of many individuals non-migratory. It could well be that the non-migratory nature of the hybrids is their most important characteristic and explains 'leap-frog' migration more satisfactorily than previously. In addition it could be the lack of migration that has resulted in the failure to assume a full breeding plumage. Redshanks nesting on the Ribble marshes which carry out a long migration have more breeding plumage than those moving only short distances and this correlation is significant at the 99 % level in over 70 sightings of wing-tagged birds and ringing recoveries for November to January inclusive. It appears that the ability to winter further north is the primary reason for success in these hybrids and results in the peculiar moult, which clearly identifies most individuals as possible hybrids. It has been suggested (Hale 1982) that at least on this occasion secondary hybridization arising from interbreeding between populations of waders from Iceland/Greenland

and Scandinavia resulted in an increased heterogeneity which produced new evolutionary potential. Throughout the Pleistocene this probably happened on a number of occasions, and has resulted in the hybrids of all the species concerned wintering on estuaries much to the north of those originally occupied by the parent stock.

Changing migration routes

Compared with that of the present day, the timberline was displaced between 2500–3000 km to the south during the height of the last glaciation. It follows that the breeding areas of at least the high-arctic breeders were similarly displaced, and though temperate breeding waders were not so much affected they were restricted to more limited breeding areas than at present. Conversely, at the height of the last interglacial, temperate breeders probably occupied a much enlarged breeding area compared with the present day and high-arctic breeders a very much smaller area.

At the height of the last interglacial it may well have been that high-arctic forms such as Knot, Sanderling and Turnstone, together with some Ringed Plovers, Dunlins and Golden Plovers, remained for the winter in Greenland. Probably the commonest waders on the European estuaries at this time were the temperate forms, the Oystercatchers (*Haematopus ostralegus*), Redshanks, Black-tailed Godwits, Curlews (*Numenius arquata*), Avocets (*Recurvirostra avosetta*) and Ruff (*Philomachus pugnax*), and, with the possible exception of Oystercatchers, their numbers were likely to have been much greater than at present. Clearly, at different phases of the interglacial, populations of waders would be very different; where grasslands were widespread, for example before the northward spread of trees, Black-tailed Godwits, Curlews and Ruff were probably present in larger numbers than after afforestation. On the other hand, Oystercatchers were probably present in large numbers throughout this period and possibly species such as the Kentish Plover (*Charadrius alexandrinus*) and even the Black-winged Stilt (*Himantopus himantopus*) may have had considerable wintering populations because of the mild climate. It is doubtful if large numbers of such species as the Grey Plover (*Squatarola squatarola*), Stints, Curlew Sandpiper (*Calidris testacea*), Spotted Redshank (*Tringa erythropus*), Greenshank (*Tringa nebularia*), Bar-tailed Godwit and Whimbrel (*Numenius phaeopus*) wintered at latitudes south of the Mediterranean, so that possibly almost the entire breeding populations of these species west of the Taimyr Peninsula wintered on the Caspian and Black Seas, or on the coast of western and southern Europe. However, the populations of these species were probably not high because of the restricted breeding habitat.

To reach a coastal wintering area it is a shorter journey for birds breeding west of the Taimyr Peninsula to fly to the Mediterranean than to the coast of the Indian Ocean. For this reason alone a migration route may well have evolved in the Ipswichian Interglacial or before to enable species such as the Curlew Sandpiper to reach a suitable wintering area. In the case of the high-arctic breeders in Greenland and on Ellesmere Island, at the present time, to follow a route into North America would necessitate a movement south of 27.5° of latitude further than moving into Europe through the British Isles, to reach a wintering area with the same January mean temperature. It is probably not coincidental that these routes are those along which the habitat was colonized by birds moving north with amelioration of the climate following the third glaciation.

After the last glaciation, birds from Europe would arrive in Greenland before populations moving north through Canada, simply because they were nearer to start with, and would move into Siberia until they met colonizers from further east. On the southward journey southerly winds over the ice cap and westerlies over the sea would assist birds moving into Europe as would the southwesterly Buran or Purga winds from Siberia. These winds almost certainly were involved in the selection and maintenance of the directions of migration at this time of the year. These ancient routes would be followed only so long as they provided the smallest achievable mortality. If another route gave rise to a smaller overall mortality then this would be selectively favoured.

With the onset of the Devensian Glaciation all populations of waders were pushed south, and whilst the numbers of arctic breeders increased, the numbers of temperate breeders decreased, and the spectrum of species wintering on the European estuaries changed in favour of increasing numbers of high-arctic forms. The wintering areas of all species were pushed south until at the height of the Devensian Glaciation the most northerly estuarine complex in western Europe emptied into the Atlantic in the region of the western approaches to the English Channel. The total wintering wader population then probably occupied a narrower latitudinal band than during the previous interglacial and species which had then wintered on the estuaries of Britain now wintered in Africa.

Obviously the location of suitable wintering areas influenced the migration routes. It has been argued elsewhere (Hale 1980) that good staging posts and wintering areas must provide both good feeding areas and good roosting facilities. Goss-Custard (1977) has shown that preferred feeding areas attract increasing numbers of birds as more move in, but at a slower rate than less preferred areas as density increases. Thus feeding birds become distributed over the feeding grounds. In the same way it is likely that the overall population becomes distributed over the total available wintering grounds. Populations from particular breeding areas favour particular wintering areas, so giving rise to a degree of 'allohiemy' which contributes to limiting gene flow between populations which, in turn, ensures the maintenance of traditional wintering areas. The establishment of such traditional wintering areas (Moreau 1972; Elliott *et al.* 1976; Evans 1976) ensures the creation of traditional migration routes which are normally only changed as a result of the wintering and breeding areas becoming unsuitable, through climatic changes, disturbance or geomorphological fluctuations.

Leap-frog migration

It is well known that different populations of the same species of birds migrate in different directions and over different distances and many waders behave in this way. Leap-frog migration, where more northerly breeding populations have the most southerly wintering areas, occurs in several species, most notably Redshanks and Ringed Plovers. Salomonsen (1954, 1955) explained this by suggesting that size was selected for in the winter range but Hale (1973, 1980, 1984) has argued that the phenomenon of leap-frog migration is best interpreted on the basis of hybrid populations producing larger individuals which are thus able to winter further north. If we accept Lack's (1954) hypothesis that migration is normally only undertaken if it produces overall more survivors than there would be by the population remaining in the breeding area, then it is easy to see why different populations of the same species behave in different ways. Clearly leap-frog migration could arise (and probably has done in several species of bird) entirely because of this. However, it could also arise through heterosis resulting from secondary hybridization, and probably this has occurred on numerous occasions as breeding wader populations have been subjected to the vicissitudes of the Pleistocene. Immediately a population begins to behave differently in the winter season, owing to either climatic or genetic influences, then this has a direct effect on the utilization of the wintering habitats and the predation on the invertebrate food supplies. In this way the waders themselves can effect changes in the estuarine habitats they occupy.

322 *W. G. Hale*

The future

The regular covering and uncovering of the flats by the tide ensures a relatively extensive and available food supply on a day-to-day basis but as estuaries mature so their utilization by wading birds changes. In the longer term, sea level changes greatly influence wader feeding areas and as some estuaries mature and are less used by waders, others come into use because of their developing mudflats. In the very long term, new estuary habitats will be created by future ice ages and earth movements, so that it is likely that in the long term, new wader wintering habitats will always become available. In the short term, man-made changes will undoubtedly affect the estuaries and their birds. It may well be that in reclaiming land in estuaries new habitat can be created and managed for the waders, outside any barrages developed, but at the present time waders seem well able to cope with changes. They have a long history of doing so and even in the very short term of their involvement in estuaries they seem to have coped well. The loss of cockle beds on Morecambe Bay following the severe winter of 1952 caused the Oystercatchers to winter elsewhere; wader flocks spending very high tides on the wing on the Cheshire Dee do so probably because of disturbance. Estuaries and other coastal feeding areas change almost daily; waders have experienced this for some 70 million years. During severe weather many waders move south and if the weather is sufficiently bad the remaining birds are selected against. Wader populations quickly change their habits to suit their changing habitat and possibly one reason why they are able to do this is because their habitat is so extensive. We are losing wetlands at a rapid rate, through draining and reclamation, and every effort must be made to maintain the habitat of the waders. We are helped in this, in the short term, by the natural modification of our estuaries, and in the longer term by sea level changes, but any reduction of estuarine habitat must place our waders in a more difficult position. It is possible that the existing populations could be accommodated in a much smaller complex of estuaries, but at present we do not know. It would well be that man's activities in changing the face of our estuaries could be harnessed to help provide an answer. Careful monitoring of such changes in estuaries and their wader populations may well provide the data from which to form an hypothesis which might be tested experimentally if threatened barrages ever materialize.

References

Bate, D. M. (1916). On a small collection of vertebrate remains from the Dar Harlam Cavern, Malta; with a note on a new species of the Genus *Cygnus*. *Proc. Zool. Soc. Lond.* **86:** 421–7.

Blanc, G. A. (1921). Grotta Romanelli. *Arch. Antrop. Etnol. Firenze.* **50:** 1–39.

Elliott, C. C. H., Waltner, M., Underhill, C. G., Pringle, J. S. S. & Dick, W. J. A. (1976). The migration system of the Curlew Sandpiper *Calidris ferruginea* in Africa. *Ostrich* **47:** 191–213.

Evans, P. R. (1976). Energy balance and optimal foraging strategies in shore birds: some implications for their distribution and movement in the non-breeding season. *Ardea* **64:** 117–39.

Frenzel, B. (1968). *Grundzüge der Pleistozönen vegetation – geschichte nord Eurasiens.* Steiner Verlag G.M.B.H., Weisbaden.

Goss-Custard, J. D. (1977). The ecology of the Wash. III. Density related behaviour and the possible effects of a loss of feeding grounds on wading birds (Charadrii). *J. appl. Ecol.* **14:** 721–39.

Hale, W. G. (1971). A revision of the taxonomy of the Redshank *Tringa totanus. Zool. J. Linn. Soc.* **50:** 199–268.

Hale, W. G. (1973). The distribution of the Redshank *Tringa totanus* in the winter range. *Zool. J. Linn. Soc.* **53:** 117–236.

Hale, W. G. (1980). *Waders.* Collins, London.

Hale, W. G. (1984). *Evolutionary trends in wading birds.* (In press.)

Lack, D. (1954). *The natural regulation of animal numbers.* Clarendon Press, Oxford.

Moreau, R. E. (1954). The main vicissitudes of the European Avifauna since the Pliocene. *Ibis* **96:** 411–31.

Moreau, R. E. (1972). *The Palaearctic-African bird migration systems.* Academic Press, New York and London.

Prater, A. J. (1981). *Estuary birds of Britain and Ireland.* T. & A. D. Poyser, Calton.

Salomonsen, F. (1954). The migration of the European Redshank (*Tringa totanus* L.) *Dansk. Ornith. For. Tiddskrift* **48:** 93–122.

Salomonsen, F. (1955). The evolutionary significance of bird migration. *Dann. Biol. Medd.* **22:** 1–62.

Schmidt, R. R., Koken, E. & Schuz, A. (1912). *Die diluviale Vorzeit Deutschlands.* Stuttgart.

Tooley, M. J. (1974). Sea level changes during the last 9000 years in north-west England. *Geogr. J.* **140**(1): 18–42.

Vaisanen, R. A. (1969). Evolution of the Ringed Plover (*Charadrius hiaticula* L.) during the last hundred years in Europe. *Ann. Acad. Sci. Fenn.* **149:** 1–90.

Index

abnormalities, anatomical, 183–4, 185, 187
Abra tenuis, 299
Accipiter gentilis (Goshawk), 130
age-related distribution
    Dunlin, 121, 160–74
    Oystercatcher, 121, 181–2, 183, 184, 185, 187
aggression, 142, 145–6, 198–9, 200
Agrostis spp., 85
Agrostis stolonifera, 91–2, 97
Ampharete acutifrons, 62–3
Anas acuta (Pintail), 86, 216, 219, 242, 246
Anas clypeata (Shoveler), 242
Anas crecca, *see* Teal
Anas penelope, *see* Wigeon
Anas platyrhynchos, *see* Mallard
Anas querquedula (Garganey), 242
Anas spp., roosting habits, 126–8
Anas strepera (Gadwall), 98, 242
Anatidae, roosting habits, 126–8
anatomical abnormalities, 183–4, 185, 187
Anser anser (Greylag Goose), 86, 97–8
Arca senilis, 295, 299, 300, 301, 303
Ardea spp., 131–2
Ardeola ibis (Cattle Egret), 132–4
Arenaria interpres, *see* Turnstone
Arenicola marina, 62–3, 156, 233, 249, 273
Aster tripolium, 85
Avicennia africana, 295
Avocet *(Recurvirostra avosetta)*
    African west coast, 278, 279, 280, 296
    Dutch Delta area, 256
    migration route changes, 319
    Wadden Sea, 216, 230, 231, 241, 243
Aythya ferina (Pochard), 242
Aythya fuligula (Tufted Duck), 108–9, 112, 114, 242
Aythya marila (Scaup), 114

Banc d'Arguin (Mauritania), 212, 293–308

basal metabolic rate (BMR), 30, 31, 302
Bathyporeia, 273
benthic fauna, *see* invertebrate prey;
    macrobenthic fauna
biomass density, 4
bird-days of use, 4, 9–10
bird density
    age ratio and, 165–6, 167
    comparisons of, 304, 305
    competition for food, 199–201
    food density and, 10, 14–15, 18–27, 60–3, 65
    measurement, 3–5
    mortality and, 122, 203, 204, 206–7
    territoriality and, 19, 23–4, 147, 149, 154, 158, 202–3
    theory of habitat distribution, 171
body weight, total, 30, 47–8, 49, 50
Branta bernicla, *see* Goose, Brent
Branta leucopsis, *see* Goose, Barnacle
breeding birds, Wadden Sea, 230, 240, 242–3
British Isles, 212–13, 261–74
Bucephala clangula (Goldeneye), 108, 110–11, 114, 215

Calidris alba, *see* Sanderling
Calidris alpina, *see* Dunlin
Calidris canutus, *see* Knot
Calidris ferruginea, *see* Sandpiper, Curlew
Calidris maritima (Purple Sandpiper), 296
Calidris minuta, *see* Stint, Little
Calidris temminckii (Temminck's Stint), 216
cannon-nets, 271
Capitella capitata, 62–3
Capitellidae, 299
Carcinus maenas (shore crab), 69–70, 195, 233, 248, 273
Cardium edule, see *Cerastoderma edule*

carnivorous birds, roosting behaviour, 126–7
carrying capacity, 3–4, 206
Cerastoderma (Cardium) edule, 62–3, 69, 194, 195, 233, 249, 273
Chaffinch (*Fringilla coelebs*), 124
Chara, 105
Charadriidae, 29
Charadrius alexandrinus, see Plover, Kentish
Charadrius hiaticula, see Plover, Ringed
Chenopodiaceae, 85
clam, *see Mya arenaria*
Clangula hyemalis (Long-tailed Duck), 109–10, 112–14, 215
cockle, *see Cerastoderma*
Columba palumbus (Woodpigeon), 129–30, 137
communal roosting, 119–20, 123–38
competition for feeding sites, 18, 198–201; *see also* dominance; interspecific competition; territoriality
coot (*Fulica atra*), 98, 242
cormorant (*Phalacrocorax carbo*), 250
Corophium, 233
Corophium volutator, 20, 273
Corvidae, European, 124–5
Corvus corax (Raven), 125, 132, 133
Corvus monedula (Jackdaw), 124–5, 128
crab, shore (*Carcinus maenas*), 69–70, 195, 248, 273
Crangon crangon, 233, 273
Crow, 125, 132, 133
Curlew (*Numenius arquata*)
  Banc d'Arguin, 296, 298, 301, 303
  British Isles, 268
  clam depletion, 69–82
  crab size selection, 248
  Danish Wadden Sea, 216
  density and prey density, 19, 21, 22–3, 65
  Dutch Delta area, 256, 257
  Dutch Wadden Sea, 242, 245, 246
  food consumption, 302
  German Wadden Sea, 230, 231, 232
  migration route changes, 319
  Morocco, 278, 280
  regurgitated pellets, 38
  Slikken van Vianen, 59–60, 65
  territoriality, 157–8
Cygnus columbianus bewickii (Bewick's Swan), 98
Cygnus cygnus (Whooper Swan), 11, 12
Cygnus olor (Mute Swan), 112

daily energy budget (DEB), 30–1
Danish Wadden Sea, 212, 214–23
Dee estuary, 266
density, *see* bird density; food density
diet, communal roosting and, 124–8

digestion of food, 9, 33, 84
Dollart, 225, 227, 234–5
dominance
  adult Dunlin, 173–4
  Oystercatcher, 188, 199–200
  roosting and, 121–2, 177–8, 188
Dotterel (*Eudromias morinellus*), 317
Dove, Barbary (*Streptopelia risoria*), 130, 131
droppings, *see* faeces
Duck, Long-tailed (*Clangula hyemalis*), 109–10, 112–14, 215
Duck, Tufted (*Aythya fuligula*), 108–9, 112, 114, 242
ducks
  dabbling, 127–8, 215, 216, 217, 244, 245
  diving, 101–15, 215, 222, 245
  see also individual species
Dunlin (*Calidris alpina*)
  age-related distribution, 121, 160–74
  Banc d'Arguin, 296, 299, 303, 304
  British Isles, 19, 267–8
  Danish Wadden Sea, 216
  Dutch Delta area, 256, 257
  Dutch Wadden Sea, 160–74, 245, 246, 247
  fat reserves, 44–6, 49, 50, 232
  food intake, 31, 36–40, 302
  German Wadden Sea, 230, 231, 232
  hybridization, 317, 318
  interspecific competition, 156
  Morocco, 276, 278, 279, 280, 281, 283–7
  night feeding, 43
  population expansion and contraction, 316, 317
  Slikken van Vianen, 59–60, 65
Dutch Delta area, 20, 57–8, 88, 253–60
Dutch Wadden Sea, 88–9, 160–74, 178–89, 212, 238–50

earthworms, 195
East Atlantic flyway, 296, 297–8
Egret, Cattle (*Ardeola ibis*), 132–4
Eider Duck (*Somateria mollissima*)
  Danish Wadden Sea, 215, 222
  dominance, 188
  Dutch Wadden Sea, 240, 241, 242, 245, 246
  Swedish west coast, 107–8, 110–12, 114
energy balance, 29–52
energy intake
  Bar-tailed Godwit, 302, 306–7
  cyclical variations, 40–2, 44
  daily, 30–1, 302
  Dunlin, 36–40
  estimation, 31, 32–40, 302
  fat reserves and, 15, 49, 51
  food density and, 9–10, 15–16, 94–5
  night, 42–4, 151–2, 153
  preferred feeding areas and, 197–8

energy requirements
    cyclical variations, 40, 44, 51–2
    estimation, 30–1
    fat reserves and, 49, 51
Enteromorpha, 9–10, 58, 89
Essex and Suffolk coasts, 21, 22–3
estuary(ies)
    configuration changes, 312–14, 322
    shape, 16
    water circulation, 105
Eudromias morinellus (Dotterel), 317
Eurydice pulchra, 273
Excirolana, 25
Exe estuary, 191–207
exposure, coastal, 105

faeces
    counts of, 86
    prey estimation, 33, 37, 38, 39
Falco columbarius (Merlin), 129
fat reserves, 15, 44–51
fat scores, 47
feeding
    night, 17, 42–4, 94–5, 149–52, 307
    social, 124–8
    time spent, 32, 40–1
feeding sites
    competition, 18, 198–201; *see also*
        dominance; interspecific competition;
        territoriality
    information sharing, 130–6
    movements between, 52
    roosting sites and, 177–8, 187
    selection, 5, 21
Festuca, 89
finches, 124
fish, 233, 300, 301, 303
fish-feeding ducks, 102, 106, 107
flock density, *see* bird density
flock size
    foraging success, 136–7
    limits on, 10
    reaction to predators and, 130
flood protection barriers, 17, 253, 254
food density
    grazing intensity and, 90–5, 98
    grazing wildfowl numbers and, 9–12, 98–9
    wader density and, 13–17, 18–27, 60–3,
        64–5
food intake, *see* energy intake
food requirements, *see* energy
    requirements
food resources
    bird movements and, 84–99
    bird numbers and, 8–27
    diving duck populations, 101–15
    wader distribution and, 57–68
    see also food density; invertebrate prey;
        macrobenthic fauna; vegetation

Forth, Firth of, 21–3
Fringilla coelebs (Chaffinch), 124
Fulica atra (Coot), 98, 242

Gadwall (*Anas strepera*), 98, 242
Gallinago gallinago (Snipe), 216, 242, 277,
    278
Gallinula chloropus (Moorhen), 242
Gammarus, 104
Garganey *Anas querquedula*, 242
German Wadden Sea, 212, 224–35
gizzard contents, 33, 38–9
glaciation, effects of, 311–17, 319–21
Godwit, Bar-tailed (*Limosa lapponica*)
    Banc d'Arguin, 296, 297, 298, 299, 303,
        304
    British Isles, 263, 268
    diet, 156, 157, 299, 303
    Dutch Delta area, 256, 257
    fat reserves, 45
    food intake, 31, 302, 306–7
    incomplete moulting, 318
    Morocco, 278, 279, 280
    population changes, 316, 319
    Slikken van Vianen, 59–60, 63, 65
    Wadden Sea, 216, 230, 231, 246
Godwit, Black-tailed (*Limosa limosa*)
    migration route changes, 319
    Morocco, 276, 278, 280, 281
    Wadden Sea, 216, 242
Goldeneye (*Bucephala clangula*), 108,
    110–11, 114, 215
Goosander (*Mergus merganser*), 244, 246
Goose, Barnacle (*Branta leucopsis*)
    droppings, 86
    Dutch Wadden Sea, 244, 246
    food consumption, 90, 95–6, 97
    food supply, 9–10
    migration routes, 86–8
    post-roost gathering, 134, 135
Goose, Brent (*Branta bernicla*)
    Danish Wadden Sea, 215, 216, 218
    Dutch Wadden Sea, 244, 246, 247
    food supply, 9–10, 11, 12
Goose, Greylag (*Anser anser*), 86, 97–8
    goose-days, 9–10
Goshawk (*Accipiter gentilis*), 130
Göta älv, river, 102, 103–4, 105, 106–7
grasses, 9, 85, 89, 93, 95, 97–8
grazing wildfowl
    food supply and numbers, 9–12
    movements and food supplies, 85–99
    roosting behaviour, 126–7
    Swedish west coast, 102, 106–7
grebes, 245
Greenshank (*Tringa nebularia*)
    Banc d'Arguin, 296, 299, 301, 303
    Dutch Delta area, 256
    food consumption, 302

migration route changes, 319–20
Morocco, 278, 280
shore crab selection, 70, 248
Slikken van Vianen, 65
Wadden Sea, 216, 230, 231
Grevelingen, 253, 255, 257, 259
Gull, Black-headed (*Larus ridibundus*),
132, 243, 246, 247
Gull, Common (*Larus canus*), 78, 243,
246
Gull, Herring (*Larus argentatus*)
crab size selection, 69–70, 248
dominance, 188
Dutch Wadden Sea, 241, 243, 246
Gull, Lesser Black-backed (*Larus fuscus*),
241, 243
gulls, 240, 241, 245
gut contents, 33, 38–9

habitat changes, adaptations to, 213,
311–21
habitat distribution, theory of, 169–74
Haematopus ostralegus, *see* Oystercatcher
Halimione portulacoides, 85
harbour seal, 250
Haringvliet, 253, 254, 255
herbivorous birds, *see* grazing wildfowl
herons, 131–2
Heteromastus filiformis, 62–3
hierarchies, *see* dominance
Himantopus himantopus (Black-winged
Stilt), 278, 280, 296, 319
human disturbance, 17, 250, 313–14; *see
also* pollution; reclamation schemes
Hydrobia ulvae, 36–40, 43, 62–3, 233,
273

Idotea pelagica, 110
information centres, communal roosts as,
130–6
insects, 233
interspecific competition, 18, 95, 143,
156–8; *see also* invertebrate prey,
partitioning of; vegetation,
complementary feeding
invertebrate prey
availability, 12–18, 25–6, 43, 127, 147–8,
150–1
density, 13–17, 18–27, 60–3, 64–5
energy content, 36, 37
estimation of intake, 34–5
identification, 33–4
measurement, 35–6
partitioning of, 18, 69–70, 72–6, 114,
248, 249–50
switching, 81
see also macrobenthic fauna

Jackdaw (*Corvus monedula*), 124–5, 128

Knot (*Calidris canutus*)
Banc d'Arguin, 296, 297, 298, 303, 304
British Isles, 266, 270
Danish Wadden Sea, 216
Dutch Delta area, 256–7, 258
Dutch Wadden Sea, 245, 246, 247
food consumption, 302
German Wadden Sea, 230, 231, 232
migration, 232, 233, 260, 298
Morocco, 278, 279, 280
population changes, 316
regurgitated pellets, 38
Slikken van Vianen, 59–60
spat predation, 79
Krammer–Volkerak, 257, 259

Lanice conchilega, 62–3
Lapwing (*Vanellus vanellus*), 216, 242, 277,
278
Larus argentatus, *see* Gull, Herring
Larus canus (Common Gull), 78, 243, 246
Larus fuscus (Lesser Black-backed Gull),
241, 243
Larus ridibundus (Black-headed Gull),
132, 243, 246, 247
Lauwerszee, 84–99
Ley Bay, 228, 234
Limosa lapponica, *see* Godwit, Bar-tailed
Limosa limosa, *see* Godwit, Black-tailed
Lindisfarne, 10–12, 14, 40–1, 42, 140,
155–8
lipid index, 45, 46–8
Littorina, 194–5, 233
Littorina littorea, 62–3
Lolium, 89
Loripes lucieus, 299
Lymnocryptes minimus, 277

Macoma balthica, 36–40, 43, 62–3, 70, 233,
249, 273
macrobenthic fauna
Banc d'Arguin, 299–301
British Isles, 273
Dutch Wadden Sea, 248–50
German Wadden Sea, 233–4
see also invertebrate prey
Mallard (*Anas platyrhynchos*), 86, 215,
216, 218, 242, 246
mangrove (*Avicennia africana*), 295
Marphysa sanguinea, 299
Mauritania, Banc d'Arguin, 212, 293–308
Melanitta fusca (Velvet Scoter), 114
Melanitta nigra (Common Scoter), 114,
215, 222
Merganser, Red-breasted (*Mergus
serrator*), 222, 244, 246
Mergus merganser (Goosander), 244, 246
Mergus serrator (Red-breasted
Merganser), 222, 244, 246

Merja Zerga, 276–7, 278, 280
Merlin (*Falco columbarius*), 129
metabolic rate, average daily, 30
migration
    lack of, 318
    leap-frog, 318, 321
migration routes
    Banc d'Arguin, 297–9
    Barnacle Goose, 86–8
    British Isles, 263–4, 265–9
    Danish Wadden Sea, 214
    Dunlin, 160–1, 166–9, 267
    Dutch Wadden Sea, 240–5, 247
    evolution of, 311–21
    food resources and, 19, 52, 84–99
    German Wadden Sea, 228–32, 233
    Goldeneye, 111
    Morocco, 276, 282–90
    Teal, 89–90
    Wigeon, 88–9
Moorhen (*Gallinula chloropus*), 242
Morecambe Bay, 261, 263, 266, 267
Morocco, Atlantic coast, 213, 276–91
mortality
    roost quality and, 121–2, 184–6, 187
    young birds, 121, 122, 172–3, 201–7
moulting
    areas, 230–2, 244–5, 257, 263
    incomplete, 318
mussels, *see Mytilus edulis*
Mya arenaria (clam), 69–82, 233, 249, 273
Mytilus edulis (Blue Mussel)
    diving ducks and, 107–8, 109–12, 114
    intertidal distribution, 58, 62–3
    nutrient supply and growth, 103, 104, 105
    Oystercatchers and, 63–8, 194, 195, 196–201
    spatfall, 69
    other mentions, 191, 192, 233, 248, 273

Nephthys hombergi, 62–3, 273
Nereis diversicolor, 22–3, 36–40, 43, 143, 157, 195, 233, 273
Nereis falsa, 299
Nereis spp., 62–3
Nereis virens, 150, 151, 152, 273
Nerine cirratulus, 273
night feeding, 17, 42–4, 94–5, 149–52, 307
Nordre älv, river, 102, 103–4, 105, 107
Notomastus latericeus, 156, 273
Numenius arquata, *see* Curlew
Numenius phaeopus, *see* Wimbrel
nutrients in coastal waters, 103–5
nutritional reserves, 15, 44–51

Oosterschelde, 57–8, 253–4, 255, 257, 258, 259

Oystercatcher (*Haematopus ostralegus*)
    Banc d'Arguin, 296, 299, 301, 303
    British Isles, 265
    clam depletion, 69–72, 73, 74, 75–82
    daily energy intake, 31, 302
    Danish Wadden Sea, 215, 216
    diet, 81, 194–6, 303
    Dutch Delta area, 256, 257, 259
    Dutch Wadden Sea, 242, 245, 246
    Exe estuary, 190–207
    feeding on mussel beds, 63–8, 196–201
    German Wadden Sea, 230, 231, 232
    Morocco, 278, 280
    mortality, 184–6, 187, 191–3, 194, 201–7
    population changes, 319
    population model, 202–7
    roosting flocks, quality of, 121–2, 177–89
    Slikken van Vianen, 59–60
    starvation, 48–9
    territoriality, 202–3

pellets, regurgitated, 33, 37–9
Phalacrocorax carbo (Cormorant), 250
Philomachus pugnax (Ruff), 216, 243, 256, 278, 280, 319
Phoca vitulina, 250
Pintail (*Anas acuta*), 86, 216, 219, 242, 246
plaice, 79
Platalea leucorodia (Spoonbill), 242
Plover, Golden (*Pluvialis apricaria*)
    Danish Wadden Sea, 216
    fat reserves, 45
    hybridization, 317, 318
    Morocco, 277, 278, 281
    speciation, 316
Plover, Grey (*Pluvialis squatarola*)
    Banc d'Arguin, 296, 299, 303, 304
    British Isles, 19, 265–6, 270
    Dutch Delta area, 256, 257
    energy content of prey, 37
    energy intake, 31, 41–2, 151, 302
    fat reserves, 45, 48
    feeding behaviour, 14, 32, 34–5, 140–58, 120–1
    migration route changes, 319–20
    Morocco, 278, 279, 280, 281
    night feeding, 17, 42, 149–52
    Slikken van Vianen, 59, 65
    territoriality, 19, 52, 120–1, 142–52, 156–8
    Wadden Sea, 216, 231, 246
Plover, Kentish (*Charadrius alexandrinus*)
    Banc d'Arguin, 296, 297, 299, 303
    Dutch Delta area, 256
    food consumption, 302
    migration route changes, 319
    Morocco, 278, 280
    Slikken van Vianen, 65
    Wadden Sea, 216, 230, 231, 241, 242

Plover, Lesser Golden (*Pluvialis dominica*), 316
Plover, Ringed (*Charadrius hiaticula*)
  Banc d'Arguin, 296, 297, 298, 303
  Dutch Delta area, 256
  energy content of prey, 37
  energy intake, 31, 40–2, 302
  feeding rate, 34
  hybridization, 318, 321
  interspecific competition, 156
  migration, 265, 321
  Morocco, 276, 278, 279, 280, 281, 283–7
  night feeding, 17, 42
  population expansion and contraction, 316–17
  Wadden Sea, 216, 230, 231, 242
Pluvialis apricaria, *see* Plover, Golden
Pluvialis dominica (Lesser Golden Plover), 316
Pluvialis squatarola, *see* Plover, Grey
Poa, 89
Pochard, (*Aythya ferina*), 242
pollution, 235, 240, 250
Polyartemia forcipata, 112
polychlorinated biphenyls, 250
population(s)
  estimates of, 3–5
  food resources and, 8–27, 101–15
  habitat changes and, 311–21
  model of Oystercatchers, 202–7
  turnover, 269–72, 282–90
Potamogeton pectinatus, 98, 105
prawns, 299, 301
predation, 128–30, 145–6
prey, *see* invertebrate prey
protein reserves, 45, 50–1
Puccinellia, 85, 93
Puerto Cansado, 277–9, 280

Quelea quelea, 134–6

Raven (*Corvus corax*), 125, 132, 133
reclamation schemes, 17, 57, 153–4, 155, 234–5
Recurvirostra avosetta, *see* Avocet
Redshank (*Tringa totanus*)
  Banc d'Arguin, 296, 298, 303
  British Isles, 268–9
  daily energy intake, 31, 302
  Dutch Delta area, 256, 257
  feeding behaviour, 32
  food density and, 19, 20, 21, 22, 65
  hybridization, 318
  migration route changes, 319, 321
  Morocco, 278, 279, 280, 281, 283–7
  nutritional reserves, 49–51
  regurgitated pellets, 38
  roosting, 128
  shore crab selection, 70, 248

Wadden Sea, 216, 230, 231, 243, 246, 247
Redshank, Spotted (*Tringa erythropus*)
  Banc d'Arguin, 296, 298
  Dutch Delta area, 256
  migration route changes, 319–20
  Morocco, 278, 280
  Slikken van Vianen, 65
  Wadden Sea, 216, 230, 231, 244
Ribble estuary, 263, 266
rodents, 112–13, 125–6
roost microclimate, 128
roosting
  communal, 119–20, 123–38
  dispersed, 136–7
  flocks, quality of, 121–2, 177–89
ruff (*Philomachus pugnax*), 216, 243, 256, 278, 280, 319
Ruppia, 105

Sahara desert, 299
Salicornia europaea, 85, 90
Salicornia spp., 85–6, 89, 90–1, 93–5, 95–6, 97
Salsola kali, 85
Sanderling (*Calidris alba*)
  Banc d'Arguin, 296, 299, 303
  British Isles, 267
  Dutch Delta area, 256
  feeding behaviour, 25
  food consumption, 302
  Morocco, 278, 280, 281
  population changes, 316
  Wadden Sea, 216, 231, 246, 247
Sandpiper, Common (*Tringa hypoleucos*), 216
Sandpiper, Curlew (*Calidris ferruginea*)
  Banc d'Arguin, 296, 297, 298–9, 303, 304
  food consumption, 302
  migration route changes, 319–20
  Morocco, 278
  Wadden Sea, 216, 230, 231
Sandpiper, Green (*Tringa ochropus*), 216
Sandpiper, Purple (*Calidris maritima*), 296
Sandpiper, Wood, 216
Scaup (*Aythya marila*), 114
Scolopacidae, 29
Scolopax rusticola (Woodcock), 242
Scoloplos armiger, 62–3
Scoter, Common (*Melanitta nigra*), 114, 215, 222
Scoter, Velvet (*Melanitta fusca*), 114
Scrobicularia plana, 62–3, 195, 273
Seal Sands, 140–54, 155, 157–8
Severn estuary, 36–40, 43
sewage discharge, 103–4
sex distribution, in roosts, 183–4

Shelduck (*Tadorna tadorna*)
  Danish Wadden Sea, 215, 216, 217, 218, 222
  Dutch Wadden Sea, 240, 242, 244, 245, 246
  population turnover, 270–1, 272
Shoveler (*Anas clypeata*), 242
Sidi Moussa–Oualidia, 279, 280, 282
sleep, 130
Slikken van Vianen, 57–68, 260
Snipe (*Gallinago gallinago*), 216, 242, 277, 278
Snipe, Jack (*Lymnocryptes minimus*), 277
Somateria mollissima, *see* Eider Duck
Spartina maritima, 295
spatfall, 69, 76, 79
speciation, 316–17
Spergularia maritima, 85
Spisula, 194, 195
Spoonbill (*Platalea leucorodia*), 242
Starling (*Sturnus vulgaris*), 128
starvation, 48–9
status signal, 173–4
Sterna albifrons (Little Tern), 243
Sterna hirundo (Common Tern), 243
Sterna paradisaea (Arctic Tern), 243
Sterna sandvicensis (Sandwich Tern), 240, 243, 244
Stilt, Black-winged (*Himantopus himantopus*), 278, 280, 296, 319
Stint, Little (*Calidris minuta*)
  Banc d'Arguin, 296, 298, 299, 303
  Danish Wadden Sea, 216
  food consumption, 302
  Morocco, 278, 279, 280
Stint, Temminck's (*Calidris temminckii*), 216
stints, 319–20
stocking density, 4
Streptopelia risoria (Barbary Dove), 130, 131
Sturnus vulgaris (Starling), 128
Suaeda maritima, 85
Swan, Bewick's (*Cygnus columbianus bewickii*), 98
Swan, Mute (*Cygnus olor*), 112
Swan, Whooper (*Cygnus cygnus*), 11, 12
Swedish west coast, 101–15

Tadorna tadorna, *see* Shelduck
Teal (*Anas crecca*)
  droppings, 86
  feeding areas, 24
  feeding at Lauwerszee, 89–90, 91–2, 95, 96
  Wadden Sea, 216, 219, 242, 246
Teesmouth, 19, 48, 140–54, 155–8, 267, 269–72

temperature
  fat reserves and, 49, 50
  prey availability and, 13, 14, 15, 147
Tern, Sandwich (*Sterna sandvicensis*), 240, 243, 244
terns, 241, 245
territoriality
  bird density and, 154, 158, 202–3
  costs and benefits, 23–4, 143–9
  Grey Plover, 19, 52, 120–1, 142–52, 156–8
  individual differences, 148
  interspecific, 18
  long-term, 141, 145–7, 151
  night, 149–52
  roosting and, 124
  short-term, 147–8, 151
Texel, 178, 179–81, 184–6
threshold density, 9–11, 98
tidal amplitude, 16–17, 105, 261
tidal levels and territoriality, 151, 152, 153
time spent feeding, 32, 40–1
topography and territory holding, 143, 144, 145, 156
Tringa erythropus, *see* Redshank, Spotted
Tringa glareola (Wood Sandpiper), 216
Tringa hypoleucos (Common Sandpiper), 216
Tringa nebularia, *see* Greenshank
Tringa ochropus (Green Sandpiper), 216
Tringa totanus, *see* Redshank
tundra refuges, 314–15, 316
Turnstone (*Arenaria interpres*)
  Banc d'Arguin, 296, 298, 303
  British Isles, 269
  Dutch Delta area, 256, 257
  food consumption, 302
  Morocco, 278, 280, 281
  population changes, 316
  Slikken van Vianen, 65
  Wadden Sea, 216, 231, 246, 247

Ulva lactuca, 105, 107

Vadehavet, *see* Wadden Sea, Danish
Vanellus vanellus (Lapwing), 216, 242, 277, 278
Vaucheria, 295
vegetation
  complementary feeding, 10–12, 95–6, 97
  Lauwerszee, 85–6, 97–8
  long-term changes, 315–16, 319
  threshold density, 9–11, 98
  see also grazing wildfowl
Vlieland, 178, 179–84, 186

Wadden Sea, 305
  Danish, 212, 214–23

Dutch, 88–9, 160–74, 178–89, 212, 238–50
  German, 212, 224–35
waders
  adaptation to habitat changes, 311–21
  Banc d'Arguin, 295–307
  British Isles, 263–9, 272–4
  Danish Wadden Sea, 215, 216, 217, 220–1
  Dutch Delta area, 255–8
  Dutch Wadden Sea, 244–5
  German Wadden Sea, 228–32
  intertidal distribution, 59–68
  Morocco, 276–82
  see also individual species
Wash, The, 25, 169, 261, 263, 266, 267, 269
water currents, Swedish west coast, 104–5
Wattenmeer, *see* Wadden Sea, German
Westerschelde, 257, 259
wheat, winter, 10
Whimbrel (*Numenius phaeopus*)
  Banc d'Arguin, 296, 297, 299, 301, 303
  food consumption, 302
  migration route changes, 319–20
  Morocco, 278, 281
  Wadden Sea, 216, 230, 231
Wigeon (*Anas penelope*)
  diet, 89, 93, 95, 98

droppings, 86
grazing *Salicornia*, 90, 92–6, 97
grazing *Zostera*, 10–12
migration routes, 88–9
Wadden Sea, 88–9, 216, 219, 245, 246
wind
  heat loss and, 14–15, 49
  prey availability and, 14–15, 143, 156, 157
  Swedish west coast, 105
wintering areas, habitat changes, 311–22
wintering species
  Banc d'Arguin, 295–8
  British Isles, 263–4
  Danish Wadden Sea, 215, 216
  Dutch Wadden Sea, 245–7
  German Wadden Sea, 231, 232
  Morocco, 276–82
Woodpigeon, 129–30, 137

Ythan estuary, 20, 169

Zostera marina, 12, 58, 105
Zostera nana, 12
Zostera noltii, 58, 295
Zostera spp., 9–12, 89